Lecture Notes in Artificial Intelligence   11330

Subseries of Lecture Notes in Computer Science

More information about this series at http://www.springer.com/series/1244

Ulf Brefeld · Jesse Davis ·
Jan Van Haaren · Albrecht Zimmermann (Eds.)

Machine Learning
and Data Mining
for Sports Analytics

5th International Workshop, MLSA 2018
Co-located with ECML/PKDD 2018
Dublin, Ireland, September 10, 2018
Proceedings

 Springer

Editors
Ulf Brefeld
Leuphana University
Lüneburg, Germany

Jan Van Haaren
SciSports
Enschede, The Netherlands

Jesse Davis 🆔
Katholieke Universiteit Leuven
Heverlee, Belgium

Albrecht Zimmermann 🆔
Université de Caen Normandie
Caen, France

ISSN 0302-9743 ISSN 1611-3349 (electronic)
Lecture Notes in Artificial Intelligence
ISBN 978-3-030-17273-2 ISBN 978-3-030-17274-9 (eBook)
https://doi.org/10.1007/978-3-030-17274-9

LNCS Sublibrary: SL7 – Artificial Intelligence

This Springer imprint is published by the registered company Springer Nature Switzerland AG
The registered company address is: Gewerbestrasse 11, 6330 Cham, Switzerland

Preface

The Machine Learning and Data Mining for Sports Analytics Workshop aims to bring people from outside of the machine learning and data mining community into contact with researchers from that community who are working in sports analytics. Editions of the workshop have been co-located with the European Conference on Machine Learning and Principles and Practice of Knowledge Discovery. The fifth edition was co-located with ECML PKDD 2018 in Dublin, Ireland.

Sports analytics has been a steadily growing and rapidly evolving area during the past decade, both in US professional sports leagues and in European football leagues. The recent implementation of strict financial fair-play regulations in European football will definitely increase the importance of sports analytics in the coming years. In addition, there is the popularity of sports betting. The developed techniques are being used for decision support in all aspects of professional sports, including:

- Match strategy, tactics, and analysis
- Player acquisition, player valuation, and team spending
- Training regimens and focus
- Injury prediction and prevention
- Performance management and prediction
- Match outcome and league table prediction
- Tournament design and scheduling
- Betting odds calculation

The interest in the topic has grown so much that there is now an annual conference on sports analytics at the MIT Sloan School of Management, which has been attended by representatives from over 70 professional sports teams in eminent leagues such as the Major League Baseball, National Basketball Association, National Football League, National Hockey League, Major League Soccer, English Premier League, and the German Bundesliga. Furthermore, sports data providers such as OPTA have started making performance data publicly available to stimulate researchers who have the skills and vision to make a difference in the sports analytics community.

The majority of techniques used in the field so far are statistical but there has been growing interest in the machine learning and data mining community in recent years, as past editions of MLSA show.

The fifth edition included a variety of topics, covering the team sports American football, basketball, ice hockey, and soccer, as well as contributions on individual sports: cycling and martial arts. In addition, there were four challenge papers, reporting on how to predict pass receivers in soccer. Moreover, the mining and learning techniques used were drawn from different research fields, demonstrating the increasing maturity of the topic that has now also begun to filter into general conference programs.

February 2018

Ulf Brefeld
Jesse Davis
Jan Van Haaren
Albrecht Zimmermann

Organization

Workshop Organizers

Ulf Brefeld	Leuphana University, Germany
Jesse Davis	Katholieke Universiteit Leuven, Belgium
Jan Van Haaren	SciScorts, Netherlands
Albrecht Zimmermann	Université de Caen Normandie, France

Program Committee

Bart Aalbers	SciSports, Netherlands
Gennady Andrienko	IAIS Fraunhofer/City University London, UK
Marcus Bendtsen	Linköping University, Sweden
Matthew van Bommel	Sacramento Kings, USA
Lotte Bransen	SciSports, Netherlands
Paolo Cintia	University of Pisa, Italy
Anthony Constantinou	Queen Mary University of London, UK
Tom Decroos	Katholieke Universiteit Leuven, Belgium
Vladimir Dzyuba	Flanders Make, Belgium
Martin Eastwood	www.pena.lt/y, UK
Johannes Fürnkranz	TU Darmstadt, Germany
Dries Goossens	Ghent University, Belgium
Derek Greene	University College Dublin, Ireland
Thomas Gärtner	University of Nottingham, UK
Mehdi Kaytoue	Infologic, France
John Komar	University of Rouen/CETAPS, France
Orestis Kostakis	Microsoft, USA
Jan Lasek	Systems Research Institute/Polish Academy of Sciences, Poland
Christophe Ley	Ghent University, Belgium
Tim Op De Beéck	Katholieke Universiteit Leuven, Belgium
Dominic Orth	Vrije Universiteit Amsterdam, Netherlands
Luca Pappalardo	University of Pisa/KDDLab, Italy
Chedy Raïssi	Institut national de recherche en informatique et en automatique, Nancy, France
François Rioult	Université de Caen Normandie, France
Oliver Schulte	Simon Fraser University, Canada
Andrew Shattock	SciSports, Netherlands
Toon Van Craenendonck	Katholieke Universiteit Leuven, Belgium
Guillermo Vinue	University of Valencia, Spain
Joshua Weissbock	University of Ottawa, Canada

Contents

Challenge Papers

Soccer

Measuring Football Players' On-the-Ball Contributions from Passes During Games

Lotte Bransen[✉] and Jan Van Haaren[✉]

SciSports, Hengelosestraat 500, 7521 AN Enschede, Netherlands
{l.bransen,j.vanhaaren}@scisports.com

Abstract. Several performance metrics for quantifying the in-game performances of individual football players have been proposed in recent years. Although the majority of the on-the-ball actions during games constitutes of passes, many of the currently available metrics focus on measuring the quality of shots only. To help bridge this gap, we propose a novel approach to measure players' on-the-ball contributions from passes during games. Our proposed approach measures the expected impact of each pass on the scoreline.

Keywords: Football analytics · Player performance · Pass values

1 Introduction

The performances of individual football players in games are hard to quantify due to the low-scoring nature of football. During major tournaments like the FIFA World Cup, the organizers[1] and mainstream sports media report basic statistics like distance covered, number of assists, number of saves, number of goal attempts, and number of completed passes [2]. While these statistics provide some insights into the performances of individual football players, they largely fail to account for the circumstances under which the actions were performed. For example, successfully completing a forward pass deep into the half of the opponent is both more difficult and more valuable than performing a backward pass on the own half without any pressure from the opponent whatsoever.

In recent years, football analytics researchers and enthusiasts alike have proposed several performance metrics for individual players. Although the majority of these metrics focuses on measuring the quality of shots, there has been an increasing interest in quantifying other types of individual player actions [4, 7, 12]. This recent focus shift has been fueled by the availability of more extensive data and the observation that shots only constitute a small portion of the on-the-ball actions that football players perform during games [12]. In the present work, we use a dataset comprising over twelve million on-the-ball actions of which only 2% are shots. Instead, the majority of the on-the-ball actions are passes (75%), dribbles (13%), and set pieces (10%).

---

[1] https://www.fifa.com/worldcup/statistics/.

© Springer Nature Switzerland AG 2019
U. Brefeld et al. (Eds.): MLSA 2018, LNAI 11330, pp. 3–15, 2019.
https://doi.org/10.1007/978-3-030-17274-9_1

In this paper, we introduce a novel approach to measure football players' on-the-ball contributions from passes during games. Our approach measures the expected impact of each pass on the scoreline. We value a pass by computing the difference between the expected reward of the possession sequence constituting the pass both before and after the pass. That is, a pass receives a positive value if the expected reward of the possession sequence after the pass is higher than the expected reward before the pass. Our approach employs a k-nearest-neighbor search with dynamic time warping (DTW) as a distance function to determine the expected reward of a possession sequence. Our empirical evaluation on an extensive real-world dataset shows that our approach is capable of identifying different types of impactful players like the ball-playing defender Ragnar Klavan (Liverpool FC), the advanced playmaker Mesut Özil (Arsenal), and the deep-lying playmaker Toni Kroos (Real Madrid).

2   Dataset

Our dataset comprises game data for 9,061 games in the English Premier League, Spanish LaLiga, German 1. Bundesliga, Italian Serie A, French Ligue Un, Belgian Pro League and Dutch Eredivisie. The dataset spans the 2014/2015 through 2017/2018 seasons and was provided by SciSports' partner Wyscout.[2] For each game, the dataset contains information on the players (i.e., name, date of birth and position) and the teams (i.e., starting lineup and substitutions) as well as play-by-play event data describing the most notable events that happened on the pitch. For each event, the dataset provides the following information: timestamp (i.e., half and time), team and player performing the event, type (e.g., pass or shot) and subtype (e.g., cross or high pass), and start and end location. Table 1 shows an excerpt from our dataset showing five consecutive passes.

Table 1. An excerpt from our dataset showing five consecutive passes.

Half	Time (s)	Team	Player	Type	Subtype	start_x	end_x	start_y	end_y
1	8.642	679	217031	8	85	58	66	34	9
1	10.167	679	86307	8	85	66	85	9	17
1	11.987	679	3443	8	85	85	90	17	25
1	13.681	679	4488	8	80	90	89	25	44
1	14.488	679	3682	8	85	89	81	44	39

3   Approach

Valuing individual actions such as passes is challenging due to the low-scoring nature of football. Since football players get only a few occasions during games

---

[2] https://www.wyscout.com.

to earn reward from their passes (i.e., each time a goal is scored), we resort to computing the passes' expected rewards instead of distributing the actual rewards from goals across the preceding passes. More specifically, we compute the number of goals expected to arise from a given pass if that pass were repeated many times. To this end, we propose a four-step approach to measure football players' expected contributions from their passes during football games. In the remainder of this section, we discuss each step in turn.

3.1 Constructing Possession Sequences

We split the event stream for each game into a set of possession sequences, which are sequences of events where the same team remains in possession of the ball. The first possession sequence in each half starts with the kick-off. The end of one possession sequence and thus also the start of the following possession sequence is marked by one the following events: a team losing possession (e.g., due to an inaccurate pass), a team scoring a goal, a team committing a foul (e.g., an offside pass), or the ball going out of play.

3.2 Labeling Possession Sequences

We label each possession sequence by computing its expected reward. When a possession sequence *does not* result in a shot, the sequence receives a value of zero. When a possession sequence *does* result in a shot, the sequence receives the expected-goals value of the shot. This value reflects how often the shot can be expected to yield a goal if the shot were repeated many times. For example, a shot having an expected-goals value of 0.13 is expected to translate into 13 goals if the shot were repeated 100 times.

Building on earlier work, we train an expected-goals model to value shots [5]. We represent each shot by its location on the pitch (i.e., x and y location), its distance to the center of the goal, and the angle between its location and the goal posts. We label the shots resulting in a goal as positive examples and all other shots as negative examples. We train a binary classification model that assigns a probability of scoring to each shot.

3.3 Valuing Passes

We split each possession sequence into a set of possession subsequences. Each subsequence starts with the same event as the original possession sequence and ends after one of the passes in that sequence. For example, a possession sequence consisting of a pass 1, a pass 2, a dribble and a pass 3 collapses into a set of three possession subsequences. The first subsequence consists of pass 1, the second subsequence consists of pass 1 and pass 2, and the third subsequence consists of pass 1, pass 2, the dribble, and pass 3.

We value a given pass by computing the difference between the expected reward of the possession subsequence after that pass and the expected reward

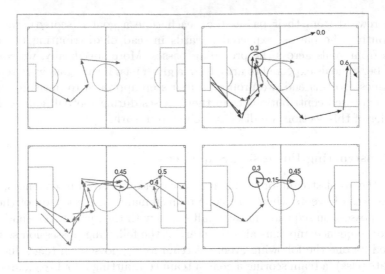

Fig. 1. Visualization of our approach to value passes, where we value the last pass of the possession sequence shown in green. We obtain a pass value of 0.15, which is the difference between the value of the possession subsequence after the pass (0.45) and the value of the possession subsequence before the pass (0.30). (Color figure online)

of the possession subsequence before that pass. Hence, the value of the pass reflects an increase or decrease in expected reward. We assume that a team can only earn reward whenever it is in possession of the ball. If the pass is the first in its possession subsequence, we set the expected reward of the possession subsequence before the pass to zero. If the pass is unsuccessful and thus marks the end of its possession subsequence, we set the expected reward of the possession subsequence after the pass to zero.

We compute the expected reward of a possession subsequence by first performing a k-nearest-neighbors search and then averaging the labels of the k most-similar possession subsequences. We use dynamic time warping (DTW) to measure the similarity between two possession subsequences [1]. We interpolate the possession subsequences and obtain the x and y coordinates at fixed one-second intervals. We first apply DTW to the x coordinates and y coordinates separately and then sum the differences in both dimensions.

To speed up the k-nearest-neighbors search, we reduce the number of computations by first clustering the possession subsequences and then performing DTW within each cluster. We divide the pitch into a grid of cells, where each cell is 15 m long and 17 m wide. Hence, a default pitch of 105 m long and 68 m wide yields a 7-by-4 grid. We represent each cluster as an origin-destination pair and thus obtain 784 clusters (i.e., 28 origins × 28 destinations). We assign each possession subsequence to exactly one cluster based on its start and end location on the pitch.

Figure 1 shows a visualization of our approach for valuing passes. In this example, we aim to value the last pass in the possession sequence shown in green (top-left figure). First, we compute the value of the possession subsequence before the pass (top-right figure). We compute the average of the labels of the two nearest neighbors, which are 0.0 and 0.6, and obtain a value of 0.3. Second, we compute the value of the possession subsequence after the pass (bottom-left figure). We compute the average of the labels of the two nearest neighbors, which are 0.4 and 0.5, and obtain a value of 0.45. Third, we compute the difference between the value after the pass and the value before the pass to obtain a pass value of 0.15 (bottom-right figure).

3.4   Rating Players

We rate a player by first summing the values of his passes for a given period of time (e.g., a game, a sequence of games or a season) and then normalizing the obtained sum per 90 min of play. We consider all types of passes, including open-play passes, goal kicks, corner kicks, and free kicks.

4   Experimental Evaluation

In this section, we present an experimental evaluation of our proposed approach. We introduce the datasets, present the methodology, investigate the impact of the parameters, and present results for the 2017/2018 season.

4.1   Datasets

We split the available data presented in Sect. 2 into three datasets: a train set, a validation set, and a test set. We respect the chronological order of the games. Our train set covers the 2014/2015 and 2015/2016 seasons, our validation set covers the 2016/2017 season, and our test set covers the 2017/2018 season. Table 2 shows the characteristics of our three datasets.

Table 2. The characteristics of our three datasets.

	Train set	Validation set	Test set
Games	4,253	2,404	2,404
Possession sequences	1,878,593	972,526	970,303
Passes	3,425,285	1,998,533	2,023,730
Shots	95,381	53,617	54,311
Goals	9,853	5,868	5,762

4.2   Methodology

We use the XGBoost algorithm to train the expected-goals model.[3] After optimizing the parameters using a grid search, we set the number of estimators to 500, the learning rate to 0.01, and the maximum tree depth to 5. We use the dynamic time warping implementation provided by the dtaidistance library to compute the distances between the possession subsequences.[4] We do not restrict the warping paths in the distance computations.

Inspired by the work from Liu and Schulte on evaluating player performances in ice hockey, we evaluate our approach by predicting the outcomes of future games as we expect our pass values to be predictors of future performances [10]. We predict the outcomes for 1,172 games in the English Premier League, Spanish LaLiga, German 1. Bundesliga, Italian Serie A and French Ligue Un. We only consider games involving teams for which player ratings are available for at least one player in each line (i.e., goalkeeper, defender, midfielder or striker).

We assume that the number of goals scored by each team in each game is Poisson distributed [11]. We use the player ratings obtained on the validation set to determine the means of the Poisson random variables representing the expected number of goals scored by the teams in the games in the test set. We compute the Poisson means by summing the ratings for the players in the starting line-up. For players who played at least 900 min in the 2016/2017 season, we consider their actual contributions. For the remaining players, we use the average contribution of the team's players in the same line. Since the average reward gained from passes (i.e., 0.07 goals per team per game) only reflects around 5% of the average reward gained during games (i.e., 1.42 goals per team per game), we transform the distribution over the total player ratings per team per game to follow a similar distribution as the average number of goals scored by each team in each game in the validation set. We compute the probabilities for a home win, a draw, and an away win using the Skellam distribution [8].

4.3   Impact of the Parameters

We now investigate how the clustering step impacts the results and what the optimal number of neighbors in the k-nearest-neighbors search is.

Impact of the Clustering Step. For an increasing number of possession sequences, performing the k-nearest-neighbors search quickly becomes prohibitively expensive. For example, obtaining results on our test set would require over 1.8 trillion distance computations (i.e., 1,878,593 possession sequences in the train set × 970,303 possession sequences in the test set). To reduce the number of distance computations, we exploit the observation that possession sequences starting or ending in entirely different locations on the pitch are unlikely to be similar. For example, a possession sequence starting in a team's penalty area is

---

[3] https://xgboost.readthedocs.io/en/latest/.
[4] https://github.com/wannesm/dtaidistance.

unlikely to be similar to a possession sequence starting in the opponent's penalty area. More specifically, as explained in Sect. 3.3, we first cluster the possession sequences according to their start and end locations and then perform the k-nearest-neighbors search within each cluster.

To evaluate the impact of the clustering step on our results, we arbitrarily sample 100 games from the train set and 50 games from the validation set. The resulting train and validation subsets consist of 68,907 sequences and 35,291 sequences, respectively. Table 3 reports the total runtimes, the number of clusters, and the average cluster size for three settings: no clustering, clustering with grid cells of 15 by 17 m, and clustering with grid cells of 5 by 4 m. As expected, clustering the possession sequences speeds up our approach considerably.

In addition, we also investigate the impact of the clustering step in a more qualitative fashion. We randomly sample three possession sequences and 100 games comprising 32,245 possession sequences from our training set. We perform a three-nearest-neighbors search in both the no-clustering setting and the clustering setting with grid cells of 15 by 17 m. Figure 2 shows the three nearest neighbors for each of the three possession sequences in both settings, where the results for the clustering setting are shown on the left and the results for the no-clustering setting are shown on the right. Although the obtained possession sequences are different, the three-nearest-neighbors search obtains highly similar neighbors in both settings.

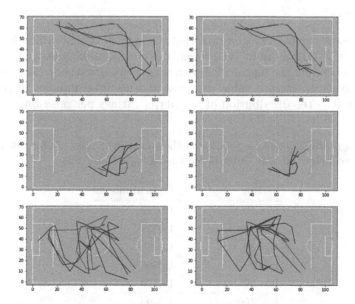

Fig. 2. Visualization of the three nearest neighbors obtained for three randomly sampled possession sequences in both the clustering setting with grid cells of 15 by 17 m and the no-clustering setting. The results for the former setting are on the left, whereas the results for the latter setting are on the right.

Table 3. The characteristics of three settings: no clustering, clustering with grid cells of 15 by 17 m, and clustering with grid cells of 5 by 4 m.

	No clustering	Cell: 15 × 17	Cell: 5 × 4
Total runtime	270 min	12 min	150 min
Number of clusters	1	784	127,449
Average cluster size	68,907	87.89	0.54

Optimal Number of Neighbors in the K-Nearest-Neighbors Search. We investigate the optimal number of neighbors in the k-nearest-neighbors search to value passes. We try the following values for the parameter k: 1, 2, 5, 10, 20, 50 and 100. We predict the outcomes of the games in the test set as explained in Sect. 4.2. Table 4 shows the logarithmic losses for each of the values for k. We find that 10 is the optimal number of neighbors.

In addition, we compare our approach to two baseline approaches. The first baseline approach is the pass accuracy. The second baseline approach is the prior distribution over the possible game outcomes, where we assign a probability of 48.42% to a home win, 28.16% to an away win, and 23.42% to a draw. Our approach outperforms both baseline approaches.

4.4   Results

We now present the players who provided the highest contributions from passes during the 2017/2018 season. We present the overall ranking as well as the top-ranked players under the age of 21. Furthermore, we investigate the relationship between a player's average value per pass and his total number of passes per 90 min as well as the distribution of the player ratings per position.

Table 4. Logarithmic losses for predicting the games in the test set for different numbers of nearest neighbors k in order of increasing logarithmic loss.

Setting	Logarithmic loss
$k = 10$	1.0521
$k = 5$	1.0521
$k = 20$	1.0528
$k = 2$	1.0560
$k = 50$	1.0579
$k = 100$	1.0594
$k = 1$	1.0725
Pass accuracy	1.0800
Prior distribution	1.0860

Following the experiments above, we set the number of nearest neighbors k to 10 and perform clustering with grid cells of 15 m by 17 m. We compute the expected reward per 90 min for the players in the 2017/2018 season (i.e., the test set) and perform the k-nearest-neighbors search to value their passes on all other seasons (i.e., the train and validation set).

Table 5 shows the top-ten-ranked players who played at least 900 min during the 2017/2018 season in the English Premier League, Spanish LaLiga, German 1. Bundesliga, Italian Serie A, and French Ligue Un. Ragnar Klavan, who is a ball-playing defender for Liverpool FC, tops our ranking with an expected contribution per 90 min of 0.1133. Furthermore, Arsenal's advanced playmaker Mesut Özil ranks second, whereas Real Madrid's deep-lying playmaker Toni Kroos ranks third.

Table 6 shows the top-five-ranked players under the age of 21 who played at least 900 min during the 2017/2018 season in Europe's top-five leagues, the Dutch Eredivisie or the Belgian Pro League. Teun Koopmeiners (AZ Alkmaar) tops our ranking with an expected contribution per 90 min of 0.0806. Furthermore, Real Madrid-loanee Martin Ødegaard (SC Heerenveen) ranks second, whereas Nikola Milenković (ACF Fiorentina) ranks third.

Table 5. The top-ten-ranked players who played at least 900 min during the 2017/2018 season in Europe's top-five leagues.

Rank	Player	Team	Contribution P90
1	Ragnar Klavan	Liverpool FC	0.1133
2	Mesut Özil	Arsenal	0.1034
3	Toni Kroos	Real Madrid	0.0943
4	Manuel Lanzini	West Ham United	0.0892
5	Joan Jordán	SD Eibar	0.0830
6	Esteban Granero	Espanyol	0.0797
7	Nuri Sahin	Borussia Dortmund	0.0796
8	Mahmoud Dahoud	Borussia Dortmund	0.0775
9	Granit Xhaka	Arsenal	0.0774
10	Faouzi Ghoulam	SSC Napoli	0.0765

Figure 3 shows whether players earn their pass contribution by performing many passes per 90 min or by performing high-value passes. The five players with the highest contribution per 90 min are highlighted in red. While Lanzini and Joan Jordán do not perform many passes per 90 min, they obtain a rather high average value per pass. The dotted line drawn through Klavan contains all points with the same contribution per 90 min as him.

Figure 4 presents a comparison between a player's pass accuracy and pass contribution per 90 min. In terms of pass accuracy, forwards rate low as they typically perform passes in more crowded areas of the pitch, while goalkeepers

Table 6. The top-five-ranked players under the age of 21 who played at least 900 min during the 2017/2018 season in Europe's top-five leagues, the Dutch Eredivisie or the Belgian Pro League.

Rank	Player	Team	Contribution P90
1	Teun Koopmeiners	AZ Alkmaar	0.0806
2	Martin Ødegaard	SC Heerenveen	0.0639
3	Nikola Milenković	ACF Fiorentina	0.0617
4	Sander Berge	KRC Genk	0.0601
5	Maximilian Wöber	Ajax	0.0599

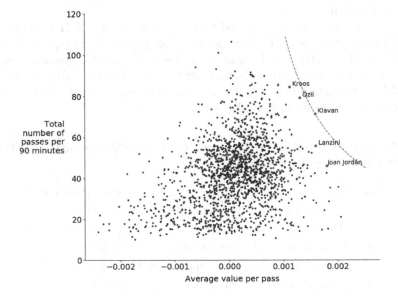

Fig. 3. Scatter plot showing the correlation between a player's average value per pass and his total number of passes per 90 min. The five players with the highest contribution per 90 min are highlighted in red. (Color figure online)

rate high. In terms of pass contribution, goalkeepers rate low, while especially midfielders rate high.

4.5   Application: Replacing Manuel Lanzini

We use our pass values to find a suitable replacement for Manuel Lanzini. The Argentine midfielder, who excelled at West Ham United throughout the 2017/2018 season, ruptured his right knee's anterior cruciate ligament while preparing for the 2018 FIFA World Cup. West Ham United are expected to sign a replacement for Lanzini, who will likely miss the entire 2018/2019 season.

To address this task, we define a "Lanzini similarity function" that accounts for a player's pass contribution per 90 min, number of passes per 90 min and pass

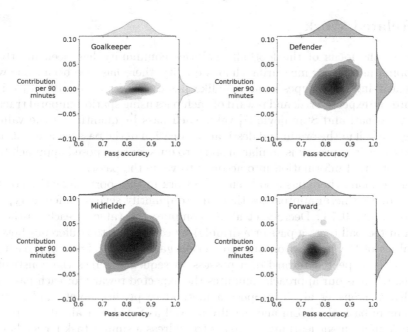

Fig. 4. Density plots per position showing the correlation between the pass accuracy and the pass contribution per 90 min.

accuracy. We normalize each of the three pass metrics before feeding them into the similarity function. Manuel Lanzini achieves a high pass contribution per 90 min despite a low pass accuracy, which suggests that the midfielder prefers high-risk, high-value passes over low-risk, low-value passes.

Table 7 shows the five most-similar players to Lanzini born after July 1st, 1993, who played at least 900 min in the 2017/2018 season. Mahmoud Dahoud (Borussia Dortmund) tops the list ahead of Joan Jordán (SD Eibar) and Naby Keïta (RB Leipzig), who moved to Liverpool during the summer of 2018.

Table 7. The five most-similar players to Manuel Lanzini born after July 1st, 1993, who played at least 900 min in the 2017/2018 season.

Rank	Player	Team	Similarity score
1	Mahmoud Dahoud	Borussia Dortmund	0.9955
2	Joan Jordán	SD Eibar	0.9881
3	Naby Keïta	RB Leipzig	0.9794
4	Dominik Kohr	Bayer 04 Leverkusen	0.9717
5	Medrán	Deportivo Alavés	0.9591

5   Related Work

Although the focus of the football analytics community has been mostly on developing metrics for measuring chance quality, there has also been some work on quantifying other types of actions like passes. Power et al. [12] objectively measure the expected risk and reward of each pass using spatio-temporal tracking data. Gyarmati and Stanojevic [7] value each pass by quantifying the value of having the ball in the origin and destination location of the pass using event data. Although their approach is similar in spirit to ours, our proposed approach takes more contextual information into account to value the passes.

Furthermore, there has also been some work in the sports analytics community on more general approaches that aim to quantify several different types of actions [3,4,6,10,13]. Decroos et al. [6] compute the value of each on-the-ball action in football (e.g., a pass or a dribble) by computing the difference between the values of the post-action and pre-action game states. Their approach distributes the expected reward of a possession sequence across the constituting actions, whereas our approach computes the expected reward for each pass individually. Cervone et al. [3] propose a method to predict points and to value decisions in basketball. Liu and Schulte as well as Schulte et al. [10,13] present a deep reinforcement learning approach to address a similar task for ice hockey.

6   Conclusion

This paper introduced a novel approach for measuring football players' on-the-ball contributions from passes using play-by-play event data collected during football games. Viewing a football game as a series of possession sequences, our approach values each pass by computing the difference between the values of its constituting possession sequence before and after the pass. To value a possession sequence, our approach combines a k-nearest-neighbor search with dynamic time warping, where the value of the possession sequence reflects its likeliness of yielding a goal.

In the future, we aim to improve our approach by accounting for the strength of the opponent and more accurately valuing the possession sequences by taking more contextual information into account. To compute the similarities between the possession sequences, we also plan to investigate spatio-temporal convolution kernels (e.g., [9]) as an alternative for dynamic time warping and to explore more sophisticated techniques for clustering the possession sequences.

References

1. Bemdt, D., Clifford, J.: Using dynamic time warping to find patterns in time series. In: Proceedings of the 3rd International Conference on Knowledge Discovery and Data Mining (1994)
2. Burnton, S.: The best passers, hardest workers and slowest players of the world cup so far. https://www.theguardian.com/football/2018/jun/29/the-best-passers-hardest-workers-and-slowest-players-of-the-world-cup-so-far

3. Cervone, D., D'Amour, A., Bornn, L., Goldsberry, K.: POINTWISE: predicting points and valuing decisions in real-time with NBA optical tracking data. In: MIT Sloan Sports Analytics Conference 2014 (2014)
4. Decroos, T., Bransen, L., Van Haaren, J., Davis, J.: Actions speak louder than goals: valuing player actions in soccer. ArXiv e-prints (2018)
5. Decroos, T., Dzyuba, V., Van Haaren, J., Davis, J.: Predicting soccer highlights from spatio-temporal match event streams. In: Proceedings of the 31st AAAI Conference on Artificial Intelligence, pp. 1302–1308 (2017)
6. Decroos, T., Van Haaren, J., Dzyuba, V., Davis, J.: STARSS: a spatio-temporal action rating system for soccer. In: Proceedings of the 4th Workshop on Machine Learning and Data Mining for Sports Analytics, vol. 1971, pp. 11–20 (2017)
7. Gyarmati, L., Stanojevic, R.: QPass: a merit-based evaluation of soccer passes. ArXiv e-prints (2016)
8. Karlis, D., Ntzoufras, I.: Bayesian modelling of football outcomes: using the Skellam's distribution for the goal difference. IMA J. Manag. Math. 20(2), 133–145 (2008)
9. Knauf, K., Brefeld, U.: Spatio-temporal convolution kernels for clustering trajectories. In: Proceedings of the Workshop on Large-Scale Sports Analytics (2014)
10. Liu, G., Schulte, O.: Deep reinforcement learning in ice hockey for context-aware player evaluation. ArXiv e-prints (2018)
11. Maher, M.: Modelling association football scores. Stat. Neerl. 36(3), 109–118 (1982)
12. Power, P., Ruiz, H., Wei, X., Lucey, P.: Not all passes are created equal: objectively measuring the risk and reward of passes in soccer from tracking data. In: Proceedings of the 23rd ACM SIGKDD International Conference on Knowledge Discovery and Data Mining, pp. 1605–1613. ACM (2017)
13. Schulte, O., Zhao, Z., Routley, K.: What is the value of an action in ice hockey? Learning a Q-function for the NHL. In: Proceedings of the 2nd Workshop on Machine Learning and Data Mining for Sports Analytics (2015)

Forecasting the FIFA World Cup – Combining Result- and Goal-Based Team Ability Parameters

Pieter Robberechts$^{(\boxtimes)}$ ⓘ and Jesse Davis ⓘ

Department of Computer Science, KU Leuven, Leuven, Belgium
{pieter.robberechts,jesse.davis}@cs.kuleuven.be

Abstract. In this study we compare result-based Elo ratings and goal-based ODM (Offense Defense Model) ratings as covariates in an ordered logit regression and bivariate Poisson model to generate predictions for the outcome of the 2018 FIFA World Cup. To this end, we first estimate probabilities of match results between all competing nations. With an evaluation on the four previous World Cups between 2002 and 2014, we show that an ordered logit model with Elo ratings as a single covariate achieves the best performance. Secondly, via Monte Carlo simulations we compute each team's probability of advancing past a given stage of the tournament. Additionally, we apply our models on the Open International Soccer Database and show that our approach leads to good predictions for domestic league football matches as well.

Keywords: Football match outcome prediction ·
Tournament simulation

1 Introduction

Association football (hereafter referred to simply as "football") is currently the most popular spectator sport in the world [25]. This popularity can be partly explained by its unpredictability [15]. Because football is such a low-scoring game, a single event can make the difference between a win, a draw or a loss. Especially on the top-level, many games are decided by an extraordinary action of a single player, a rare defensive slip, a refereeing error, or just luck. As a consequence, many football games are ultimately won by the proclaimed underdog.

Despite the fact that a large part of the outcome of soccer matches is governed by chance, every team has its strengths and weaknesses and most results reflect these qualities. Eventually, skill does prevail and the best teams typically distinguish themselves during the course of a season or a tournament. This indicates that statistical forecasting methods could be used to predict the outcome of football matches.

Making these predictions is also one of the favorite pastimes of many football fans. However, they typically base their predictions on subjective opinions. It is

U. Brefeld et al. (Eds.): MLSA 2018, LNAI 11330, pp. 16–30, 2019.
https://doi.org/10.1007/978-3-030-17274-9_2

a challenging task to quantify the strength of a team objectively. There has previously been a fair amount of research on this topic. The Elo rating system is one of the most adopted approaches. While its origins are in chess [10], Elo ratings are commonly used for other sports, including football. Hvattum and Arntzen [24] have shown for English league football that an ordered logit model with the relative difference between the Elo ratings of two competing teams as a single covariate is a highly significant predictor of match outcomes. While the Elo system is essentially a result-based rating (i.e., Elo ratings are computed from the win-draw-loss records of a team), other rating systems are goal-based (i.e., they are based on the number of goals a team scores). These approaches typically extend ratings to two parameters – an offensive and defensive rating.

In this paper we compare both rating systems (as well as a combination of both) in terms of their predictive performance on previous World Cups and we provide our predictions for the 2018 World Cup. Our approach consists of three steps: First, we estimate the strength of the participating teams using past results. Second, we estimate probabilities of match results between each competing nation based on the pairwise rating differences. Finally, the predictions are used to determine the most likely tournament outcome in a Monte Carlo simulation.

The remainder of this paper is structured as follows: In Sect. 2, we discuss various proposals that have been made for modeling the outcome of football matches, as well as for rating the strength of teams; Next, Sect. 3, describes our models, followed in Sect. 4 by a discussion of three different performance metrics to compare the predictive power of these models. Finally, Sects. 5 and 6 validate the predictive strength of our models on respectively international and domestic league football.

2   Related Work

The literature on modeling the outcomes of football games can be divided in two broad categories, namely goal-based and result-based models. The first category models the number of goals scored and conceded by both competing teams. Predictions of win-draw-loss outcomes can then be derived indirectly by aggregating the probabilities assigned to all possible scorelines. The second category models win-draw-loss outcomes directly.

The simplest goal-based models assume that the number of goals scored by both teams are independent and can be modeled with two separate models. Poisson regression models are used predominately, but the negative binomial distribution [2] and Gaussian distribution (by employing the least squares regression method) [40] appear as well in the literature. For example, Lee et al. [31] applied a Poisson regression model to data from the 95/96 Premier League season, using the offensive and defensive strength of both teams and the home advantage as parameters. These parameters are then estimated using maximum-likelihood estimation on historic data. Although these independent Poisson models give a reasonably accurate description of football scores, they tend to underestimate the proportion of draws. Maher et al. [34] were the first to identify a slight correlation between scores and therefore proposed bivariate Poisson models as an alternative. They showed that such a bivariate Poisson distribution gives a better fit on differences in scores than an independent model. Yet, they did not use this insight to predict the results of future matches. Dixon and Coles [7] made the same observation, but instead of using a bivariate Poisson model, they extended the independent Poisson model by introducing an ad-hoc adjustment to the probabilities of low-scoring games. Karlis and Ntzoufras [26] further developed the idea of bivariate Poisson distributions for forecasting football games.

Approaches that model the match outcome directly are a more recent development. Apart from their computational simplicity, these models have the advantage that they avoid the problem of having to model the interdependence between the scores of both teams [17]. Most studies in this category use discrete choice regression models, such as the ordered probit model and the ordered logit model [18,28]. Goddard [17] compared a bivariate Poisson regression models with an ordered probit regression model and found that a hybrid model in which goal-based team performance covariates are used to forecast win-draw-loss match results yielded the best performance.

Although regression models are most common in the literature, any machine learning model could be used. For example, Groll et al. [21] found that a random forests model generally outperforms the conventional regression methods. Another popular class of models are Bayesian networks. Rue and Salvesen [38] proposed a Dynamic Bayesian Network (DBN) in order to take the time-dependent strength parameters of all teams in a league simultaneously into account. Baio and Blangiardo [1] extended this to a hierarchical goal-based Bayesian model. As such, they avoid the use of a more complex bivariate structure for the number of goals scored by assuming a common distribution at a higher level.

While the above studies focus on the actual prediction models, other studies have investigated the feasibility of possible covariates. Bookmaker odds are a first popular covariate. They reflect the (expert) predictions of bookmakers [37], who have strong economic incentives to make accurate predictions. Several studies have found that they are an efficient forecasting instrument [4,14,39].

A second popular covariate are ratings or rankings. The main idea is to estimate adequate ability parameters that reflect a team's current strength, based on a set of recent matches. A widely accepted approach in sports forecasting is the Elo rating system [10]. It has several generalisations, including the Glickman [16] and TrueSkill [22] rating systems. Besides Elo, there are numerous other approaches that fit into many categories [29]. Closely related to the regression-based forecasting models are the regression-based ranking methods. These models use maximum likelihood estimation to find adequate strength parameters for each team that can explain the number of goals scored or the win-tie-loss outcome in past games [33]. Other approaches are the Markov Chain based models such as Keener et al. [27] and the Power Rank rating system [45]; or the network based rating systems, such as the one by Park and Newman [36]. Yet another one is the Sinkhorn-Knopp based ranking models such as the Offense Defense Model (ODM) [19]. Lasek et al. [30] found that these Elo ratings outperform several of these other ranking methods when predicting the outcome of individual games. Additionally, Van Haaren and Davis [43] found that Elo ratings perform well when predicting the final league tables of domestic football.

3   The Models

In this section, we describe the different components of our approach for predicting the outcome of football games. We computed both result-based Elo ratings and goal-based ODM ratings for each team based on past results. These ratings are then combined in an ordered logit regression or bivariate Poisson regression model in order to make predictions for future games. In the next sections, we compare various combinations of these rating systems and regression models.

3.1   The Elo Rating System

We will briefly introduce both the basic Elo rating system and two football-specific modifications. An Elo rating system assigns a single number to each team that corresponds to a team's current strength. These numbers increase or decrease depending on the outcome of games and the ratings of the opponents in these games. Therefore, the Elo system defines an expected score for each team in a game, based on the rating difference with the opponent. Let R^H be the current rating of the home team and R^A the current rating of the away team, the exact formulae for the expected score E^H and actual score S^H of the home team are given by:

$$E^H = \frac{1}{1 + c^{R^H - R^A}/d} \quad \text{and} \quad S^H = \begin{cases} 1 & \text{if the home team won} \\ 0.5 & \text{if the match ended in a draw} \\ 0 & \text{otherwise} \end{cases}$$

The expected score and actual score for the away team are then respectively $E^A = 1 - E^H$ and $S^A = 1 - S^H$.

When a team's actual score exceeds its expected score, this is seen as evidence that a team's current rating is too low and needs to be adjusted upward. Similarly, when a team's actual score is below its expected score, that team's rating is adjusted downward. Elo's original suggestion, which is still widely used, was a simple linear adjustment proportional to the amount by which a team overperformed or underperformed their expected score. The formula for updating the rating is

$$R'^H = R^H + k(S^H - E^H).$$

How much a team's rating increases or decreases is determined by both its expected score and the k-factor. The rating of a team that was expected to win by a large margin will therefore decrease with an accordingly large amount if it actually loses. The k-factor is often called the recentness factor, because it determines how much weight is given to the results of recent matches. We added two additional factors: the competitiveness factor and the margin of victory. First, one of the difficulties in evaluating international football is that not all games are handled with the same seriousness. Friendlies, for example, are often used to experiment with new line-ups and players tend not to go to any extreme. Therefore, when computing the Elo ratings, we weight games differently depending on the importance of the competition. Second, because the best performing international teams play most of their matches against weak opponents (especially in European qualifiers) and record very few losses, we take the margin of victory (i.e., the absolute goal difference) into account [24]. So, we replace k by the expression

$$k = k_0 w_i (1 + \delta)^\gamma \tag{1}$$

with δ the absolute goal difference, $w_i > 0$ a weight factor corresponding to the competitiveness of the competition, k_0 the recentness factor and $\gamma > 0$ a fixed parameter determining the impact of the margin of victory on the update rule.

There are five parameters in this rating system. The parameters c and d determine the scale of the ratings. We set them respectively to 10 and 400. Other values lead to the same rating system, but one has to determine matching weight parameters k_0, w and γ. The optimal values for k_0 and γ are determined from historical data. We explain this procedure in Sect. 3.3. The values for w are application dependent and based on expert knowledge.

3.2   Offense-Defense Ratings

An important factor in football games is the playing style of both teams and the balance between offense and defense. We argue that, besides the relative strength of both teams, this difference in playing style might be an important factor in deciding the final outcome of a game. For example, a game between two teams that are known to rely on a very strong defense might have a higher probability to end up in a draw than a game between two teams that are known to play very offensively.

A rating system that can capture these differences in offensive and defensive strengths of a team is the Offense Defense Model (ODM) [19]. As opposed to the Elo system discussed above, it captures the offensive and defensive strength of a team as two separate parameters. Therefore, it uses goals scored as a measure of offensive strength and goals conceded as a measure of defensive strength. Whether a game eventually results in a win, a tie or a loss does not affect the ratings. We will again briefly discuss the basic ODM rating system, followed by a couple of modifications for its application to international football.

Define A_{ij} as the goals scored by team j against team i. The offensive and defensive ratings of team j are

$$o_j = \sum_{i=1}^n \frac{A_{ij}}{d_i} \quad \text{and} \quad d_j = \sum_{i=1}^n \frac{A_{ji}}{o_i}.$$

Since the offensive and defensive ratings are interdependent, they must be approximated by an iterative refinement procedure. We refer to the original paper for details.

This approach works for domestic league football where each team plays the same number of games against every other team. In international football, however, there can be large disparities between the number of games played and the strength of the opponents. A team that plays few games against strong opponents will likely score fewer goals and concede more goals, which leads to weaker attack and weaker defence ratings. To address these problems, we update ratings sequentially. In each game, a team has two sets of ratings: pre-game ratings and post-game ratings. The pre-game ratings are a weighted sum of a team's post-game ratings in previous games. Similar to what we did for the Elo rating system, these weights are determined by the recentness of a game and its competitiveness. To compute the post-game ratings, we apply the iterative procedure from the original ODM as if the competition has only two teams and using the pre-game ratings as initial estimates for the offensive and defensive ratings. Algorithm 1 below defines the exact procedure.

Algorithm 1. Computing post-game ratings from pre-game ratings

1: **procedure** SCALE(A, x)
2:     $y \leftarrow A\frac{1}{x}$
3:     $x_{\text{post}} \leftarrow A^T \frac{1}{y}$
4:     $y_{\text{post}} \leftarrow A\frac{1}{x}$
5:     **return** $x_{\text{post}}, y_{\text{post}}$
6: **end procedure**

7: $A_{ij} \leftarrow$ score team j generated against team i
8: $o_{\text{post}} \leftarrow [o_1, o_2]$ ▷ The pre-game offensive ratings of both teams
9: $d_{\text{post}} \leftarrow [d_1, d_2]$ ▷ The pre-game defensive ratings of both teams
10: **for** k=0,...,nb_iter **do**
11:     $o_{rel_o}, d_{rel_o} \leftarrow$ scale$(A,\ o_{\text{post}})$
12:     $d_{rel_d}, o_{rel_d} \leftarrow$ scale$(A^T,\ d_{\text{post}})$
13:     $o_{\text{post}} \leftarrow o_{rel_o} + o_{rel_d}$
14:     $d_{\text{post}} \leftarrow d_{rel_o} + d_{rel_d}$
15: **end for**

3.3   Match Result Predictions

The rating systems defined above can be combined with a regression model to obtain predictions for future matches. Therefore, we consider the rating differences $R^H - R^A$, $o^H - d^A$ and $d^H - o^A$ as covariates. Additionally, a fourth covariate indicates whether a home advantage applies to the home team. In Sect. 5, we compare the predictive power of an ordered logit regression model [35] and a bivariate Poisson regression model [26], as well as various combinations of these covariates in terms of their predictive power on previous World Cups.

We use the L-BFGS-B algorithm [5] with the Ranked Probability Score (RPS) [11] as a loss function to determine the optimal set of parameters for these models. This approach allows us to jointly optimize the parameters for both the rating systems and regression models. Therefore, we order the games in our dataset chronologically and define two subsets: a validation set and a test set. The test set contains the matches that we would like to predict and ideally the validation set contains matches from previous editions of the same tournament or league. Then we repeatedly evaluate each match in the complete dataset sequentially, updating the ratings based on the actual outcome and making a prediction for the matches in the validation set. Once all matches are evaluated, we compute the RPS on the validation set and update the parameters for both the rating system and regression model in order to minimize the RPS.

4 Evaluation Procedures

In this section, we consider several evaluation measures to compare our models with each other and to the odds determined by bookmakers. These odds can serve as a natural benchmark for our models. In contrast to the methods above, bookmaker odds are not solely based on results in past games. They include expert judgments from the bookmakers, which have a strong economic motivation to rate the competitors accurately [32]. After removing the profit margin of the bookmaker, the inverted odds can be interpreted as outcome probabilities [20].

Both our models and the bookmaker odds have in common that they assign a probability to all three possible outcomes of a match. One can evaluate these probabilities in three ways: First, one can consider the outcome with the highest assigned probability as the predicted outcome. Second, one can look at the probability that was assigned to the true outcome. As a third evaluation, one can judge these three probabilities as a whole. Each of these evaluations leads to a different evaluation measure, which we define below.

In the following paragraphs, we use the ordered vector $\hat{p} = (p_1, p_2, p_3)$ to denote a probability forecast of all possible match outcomes $r = (\text{win}, \text{tie}, \text{loss})$. Additionally, $y = (y_1, y_2, y_3)$ denotes the true outcome of a match, with y_i a binary indicator of whether or not i is the true outcome.

Accuracy. This measure compares the outcome with the highest assigned probability to the true outcome. The accuracy for a single match is computed using the following formula:

$$\mathbb{1}[\arg \max_i y_i = \arg \max_i \hat{p}_i] \tag{2}$$

where $\mathbb{1}[.]$ is the indicator function that equals 1 if the statement between brackets holds, and 0 otherwise.

Logarithmic loss. This measure the uncertainty of the prediction based on how much it varies from the actual outcome. The logarithmic loss is computed as

$$-\sum_{i=1}^{|r|} y_i \log \hat{p}_i \tag{3}$$

with $|r|$ the number of possible outcomes. A perfect classifier would have a logarithmic loss of precisely zero. Less ideal classifiers have progressively larger values.

Ranked Probability Score (RPS). The ranked probability score (RPS) was introduced in 1969 by Epstein [11] to evaluate probability forecasts of ranked categories. In contrast to the two previous measures it explicitly accounts for the ordinal structure of the predictions. This means that predicting a tie when the actual outcome is a loss is considered a better prediction than a win. For our purpose, it can be defined as

$$\frac{1}{|r|-1} \sum_{k=1}^{|r|-1} (\sum_{l=1}^{k} (\hat{p}_l - y_l))^2 \tag{4}$$

As the RPS is an error measure, a lower value corresponds to a better fit.

We will use the same metrics to evaluate the prediction of a tournament outcome. For that purpose, we define the set of possible outcomes as $r =$ (elimination in the group stage, elimination in the round of 16, . . . , win).

5   Validation on Previous World Cups

This section evaluates the predictive performance of our models on the four FIFA World Cups between 2002 and 2014. Therefore, we adopt a leave-one-out procedure where we iteratively tune the parameters of our model on three out of the four world cups and evaluate the performance on the left out one. For example, while predicting the matches of the 2010 World Cup (i.e., the test set), we use all available international games played between the end of the second world war and the start of the 2018 World Cup to determine the ratings for each team; but only matches of the 2002, 2004 and 2014 World Cups to tune the parameters (i.e., the validation set). Although we use data from future World Cups to determine the parameter settings for a previous World Cup, note that we only include information from past games when rating teams. Therefore, the predictions do not depend on results in future matches.

Our dataset was scraped from http://eloratings.net and includes all international games played between the end of the second world war (January 1, 1946) and the start of the 2018 World Cup (June 13, 2018). For each of these games, we scraped the competing teams, the date of the game, the competition, the outcome after the official game time, the outcome after extensions or penalties, and whether a home advantage applies. Additionally, we scraped the average assigned three-way odds by multiple bookmakers from http://betexplorer.com for all[1] World Cup matches between 2002 and 2014.

We classified all international competitions into three categories, corresponding with their competitiveness and relative importance. Each category is assigned a weight, which corresponds to w in Formula 1. We assigned a "very high" weight ($w = 1$) to World Cup games; a "high" weight ($w = 0.833$) to each of the six (AFC, CAF, CONCACAF, CONMEBOL, OFC and UEFA) continental championships; a "medium" weight ($w = 0.66$) to their qualifiers as well as to the World Cup qualifiers; a "low" weight ($w = 0.5$) to the less important tournaments, such as the African Games, Balkan Cup,...; and finally a "very low" weight ($w = 0.33$) to Friendlies These five categories originate from the FIFA/Coca-Cola World Ranking. In comparison, we added an additional category of very low importance for friendlies (the FIFA ranking gives friendlies the same weight as small tournament games) and used different weights.

Table 1 shows the predictive performance of the models introduced in Sect. 3 on the individual games from these four World Cups. Additionally, we include the averaged bookmaker predictions and the predictions from the best-performing random forest model from Groll et al. [21] as a baseline. This random forest model includes amongst others the bookmaker odds and Elo ratings as covariates. We updated the ratings of each team after each game based on the true outcome. This approach is in line with how bookmaker odds are updated until close before the start of a game. It turns out that the simple ordered logit regression model with the Elo rating difference and home advantage as covariates outperforms the bookmaker predictions and all other models in terms of accuracy, logarithmic loss and RPS. Furthermore, we notice that the 2006 and 2014 are a lot easier to predict than the 2002 and 2010 World Cups. The ODM-based models perform slightly better on these hard to predict World Cups.

Table 1. Validation of different predictive models on individual games in the 2002, 2006, 2010 and 2014 World Cups. The models are ordered by increasing average RPS value. The last column shows the average RPS for each World Cup.

Team	Accuracy	Log loss	RPS	avg ● 2002 ■ 2006 ▲ 2010 ♦ 2014
ELO ordered logit	0.5938	0.9375	0.1860	
ELO bivariate Poisson	0.5898	1.0045	0.1866	
Random forest [21]	0.5560		0.1870	
Bookmakers	0.5240	0.9594	0.1877	
ELO+ODM ordered logit	0.5820	0.9437	0.1878	
ELO+ODM bivariate Poisson	0.5703	0.9637	0.1917	
ODM ordered logit	0.5586	0.9657	0.1949	
ODM bivariate Poisson	0.5664	0.9755	0.1968	

---

[1] The odds of six games were missing.

Next, we used these models to predict the tournament course for the World Cups between 2002 and 2014. Given match outcome probabilities for each possible match-up, we ran 20,000 Monte Carlo simulations for each World Cup. Occasionally two or more teams will finish the group phase on the same points tally. In that case, the FIFA defines a couple of tie-breakers to determine the ranking order, primarily based on the number of goals scored. However, since the ordered logit models can only sample win, draw and loss results, we resolve these ties randomly. Similarly, in the case of a draw in the knockout stage, we simulate the extra time by sampling a second result. The bivariate Poisson models make it possible to resolve these equal point tallies according to the official rules, but this does not lead to more accurate predictions.

Table 2 presents the performance of the Elo-based and Elo+ODM-based ordered logit models on the past World Cups. In contrast to the previous experiment, these are pre-tournament predictions, meaning that only games preceding the corresponding World Cup are considered when making these predictions. For comparison, we added the 2014 World Cup predictions by FiveThirtyEight [12]. The model using the Elo ratings as a single covariate is again the best performing one, but it turns out that these tournament forecasts are quite inaccurate. We could predict the actual round of elimination correctly for only about half of the participating teams.

Table 2. Performance of a Monte Carlo simulation on previous World Cups, using the Elo and Elo+ODM-based ordered logit models. Additionally, we provide the 2014 World Cup predictions of FiveThirtyEight [12] for comparison.

World Cup	Model	Accuracy	Log loss	RPS	Correct by round[a]
2014	Elo	0.5000	0.2020	0.1364	10/5/3/1/0
	Elo+ODM	0.4375	0.2075	0.1381	10/4/3/1/0
	FiveThirtyEight	0.4063	0.1947	0.1334	8/6/3/1/0
2010	Elo	0.5625	0.1945	0.1330	12/4/2/1/1
	Elo+ODM	0.5313	0.1878	0.1324	11/5/2/1/1
2006	Elo	0.5313	0.1902	0.1317	12/6/2/0/0
	Elo+ODM	0.4063	0.2125	0.1422	11/5/1/1/0
2002	Elo	0.5000	0.1962	0.2322	11/4/0/0/0
	Elo+ODM	0.4688	0.2471	0.1456	10/4/0/0/0

[a] Number of teams correctly picked to advance to the next round as respectively *round of 16/quarter final/semi final/final/win*. For example, *11/4/3/1/0* means that 11 of the 16 teams advancing to the round of 16 where predicted.

Finally, we applied the ordered logit model with both Elo and ODM ratings as covariates to forecast the 2018 World Cup. According to our simulation, Brazil was the clear favorite with a win probability of 33% followed by Germany, Spain, France and Argentina. These results were in line with the bookmaker odds, although the bookmakers were more conservative about the win probability of Brazil. Table 3 shows these probabilities for the five favourites. For a detailed overview, we refer to our interactive at https://dtai.cs.kuleuven.be/sports/worldcup18/.

Table 3. Estimated probabilities for reaching the different stages in the 2018 World Cup for the five most-likely winners based on 20,000 simulations.

Team	Sixteen	Quarter	Semi	Final	Win	Bookmakers
Brazil	92%	73%	59%	46%	33%	15%
Germany	81%	57%	41%	28%	15%	15%
Spain	81%	70%	46%	26%	14%	12%
France	79%	58%	41%	21%	11%	12%
Argentina	77%	51%	32%	16%	7%	8%

The 2018 World Cup caught the attention of several other data scientist, trying to forecast the tournament outcome. Based on their SPI rating system, FiveThirtyEight [13] forecasted Brazil (19%) to win the World Cup, followed by Spain (17%) and Germany (13%). The same teams were determined as the major favorites by Zeileis, Leitner and Hornik [44]. By aggregating the winning odds of several bookmakers and transforming those into winning probabilities, they obtained a win probability of 16.6% for Brazil, 15.8% for Germany and 12.5% for Spain. The Swiss bank UBS [42] came up with the same three favorites, but in a different order. They obtained Germany as the main favorite (24.0%), followed by Brazil (19.8%) and Spain (12.5%). Also Groll et al. [21] came up with Spain (17.8%), Germany (17.1%) and Brazil (12.3%) as the main favorites. They combined a large set of 16 features with a random forest approach. As far as we know, only EA Sports [9] correctly predicted France as the World Cup winner. Yet, they did not publish any win probabilities. In Table 4 we compare these models with ours, looking both at predictive accuracy for individual games (if available) and the accuracy of the pre-tournament simulation. To allow comparison with FiveThirtyEight's predictions, we convert the win-tie-loss probabilities for games in the knockout stage to win-loss probabilities. Therefore, we use the formula $p'_{win} = p_{win} + p_{win}/(p_{win} + p_{loss}) * p_{tie}$ and analogous for p'_{loss}.

6   Validation on Domestic League Football

We also verified our models on The Open International Soccer Database that was provided as part of the 2017 Soccer Prediction Challenge [8]. The training set incorporates 216,743 match outcomes, with missing data (as part of the challenge), from 52 football leagues from all over the world in the seasons ranging from 2000–2001 until 2017–2018. The challenge involved using a single model to predict 206 future match outcomes from 26 different leagues. We used the 2010–2011 until 2017–2018 seasons of the training set as our test set to determine

Table 4. A comparison of our model's predictions with others for the 2018 World Cup. Both the predictive accuracy of individual games (if available) and of the pre-tournament forecast are listed.

	Individual games			Tournament simulation			
	Accuracy	Log loss	RPS	Accuracy	Log loss	RPS	Correct by round[a]
Bookmakers	0.6094	0.8463	**0.1976**				
FiveThirtyEight [13]	**0.6250**	**0.8457**	0.1976	0.5313	**0.1822**	**0.1242**	14/4/1/0/0
Zeileis et al. [44]				0.5625	0.1845	0.1269	14/4/1/0/0
Groll et al. [21]				**0.5938**	0.1855	0.1261	**14/4/2/0/0**
ELO+ODM logit	0.6094	0.8865	0.2072	0.5625	0.2243	0.1315	14/4/1/0/0
UBS [42]				0.5000	0.2009	0.1923	13/4/1/0/0

[a] Number of teams correctly picked to advance to the next round as respectively *round of 16/quarter final/semi final/final/win*. For example, *11/4/3/1/0* means that 11 of the 16 teams advancing to the round of 16 where predicted.

the optimal parameters for our models. This corresponds to about half of the training data.

Table 5 compares the performance of our best-performing models to the four best results of the competition and the bookmaker odds. As for the World Cup predictions, these bookmaker odds are the average assigned three-way odds by multiple bookmakers scraped from http://betexplorer.com. Although we did not optimize our approach for domestic league football, it was found that our relatively simple model outperforms all other, more complex models in terms of RPS. Three models report a better accuracy. While we did not verify this, we think that the predictive accuracy could be improved by incorporating league-specific home advantages and (because of transfers) allowing faster rating updates after the summer and winter breaks.

Table 5. Our approach compared to the best performing models from the 2017 Soccer Prediction Challenge [8]. We omit the logarithmic loss, because we do not have it for the 2017 Soccer Prediction Challenge submissions.

Model	RPS		Acc	
Bookmakers	0.2020	⊢	0.5194	⊢———⊣
ELO ordered logit	0.2035	⊢——⊣	0.5146	⊣
ELO+ODM ordered logit	0.2045	⊢———⊣	0.5146	⊣
Berrar et al. [3]	0.2054	⊢————⊣	0.5194	⊢———⊣
Hubáček et al. [23]	0.2063	⊢—————⊣	0.5243	⊢————⊣
Constantinou [6]	0.2083	⊢———————⊣	0.5146	⊣
Tsokos et al. [41]	0.2087	⊢———————⊣	0.5388	⊢——————————⊣

7 Conclusion

In this work, we compared several models for match outcome prediction and tournament simulation in football. We considered all possible combinations of a result- and goal-based regression model, with result-based Elo rating and goal-based ODM rating differences as covariates. In conclusion, we found that a very basic Elo-based ordered logit model outperforms all other models, including more complex models from the literature. The outcome of any match is unpredictable enough to confound these sophisticated computer models.

Acknowledgements. PR is supported by Interreg V A project NANO4Sports. JD is partially supported by KU Leuven Research Fund (C14/17/070 and C22/15/015), FWO-Vlaanderen (SBO-150033) and Interreg V A project NANO4Sports.

References

1. Baio, G., Blangiardo, M.: Bayesian hierarchical model for the prediction of football results. J. Appl. Stat. **37**(2), 253–264 (2010). https://doi.org/10.1080/02664760802684177
2. Baxter, M., Stevenson, R.: Discriminating between the poisson and negative binomial distributions: an application to goal scoring in association football. J. Appl. Stat. **15**(3), 347–354 (1988). https://doi.org/10.1080/02664768800000045
3. Berrar, D., Dubitzky, W., Lopes, P.: Incorporating domain knowledge in machine learning for soccer outcome prediction. Mach. Learn. **108**(1), 97–126 (2019)
4. Boulier, B.L., Stekler, H.O.: Predicting the outcomes of National Football League games. Int. J. Forecast. **19**(2), 257–270 (2003). https://doi.org/10.1016/S0169-2070(01)00144-3
5. Byrd, R.H., Lu, P., Nocedal, J., Zhu, C.: A limited memory algorithm for bound constrained optimization. SISC **16**(5), 1190–1208 (1995). https://doi.org/10.1137/0916069
6. Constantinou, A.C.: Dolores: a model that predicts football match outcomes from all over the world. Mach. Learn. **108**(1), 49–75 (2018). https://doi.org/10.1007/s10994-018-5703-7
7. Dixon, M.J., Coles, S.G.: Modelling association football scores and inefficiencies in the football betting market. J. Royal Stat. Soc. Ser. C (Appl. Stat.) **46**(2), 265–280 (1997)
8. Dubitzky, W., Lopes, P., Davis, J., Berrar, D.: The open international soccer database for machine learning. Mach. Learn. **108**(1), 9–28 (2019)
9. EA Sports: EA Sport predicts France to win the FIFA World Cup, May 2018. https://www.easports.com/fifa/news/2018/ea-sports-predicts-world-cup-fifa-18
10. Elo, A.E.: The Rating of Chess Players, Past and Present. Arco Pub., New York (1978)
11. Epstein, E.S.: A scoring system for probability forecasts of ranked categories. J. Appl. Meteorol. **8**(6), 985–987 (1969). https://doi.org/10.1175/1520-0450(1969)008<0985:ASSFPF>2.0.CO;2
12. FiveThirtyEight: 2014 World Cup Predictions, June 2014. https://fivethirtyeight.com/interactives/world-cup/
13. FiveThirtyEight: 2018 World Cup Predictions, June 2018. https://projects.fivethirtyeight.com/2018-world-cup-predictions/
14. Forrest, D., Goddard, J., Simmons, R.: Odds-setters as forecasters: the case of English football. Int. J. Forecast. **21**(3), 551–564 (2005). https://doi.org/10.1016/j.ijforecast.2005.03.003

15. Forrest, D., Simmons, R.: Outcome uncertainty and attendance demand in sport: the case of English soccer. J. Royal Stat. Soc. **51**(2), 229–241 (2002). https://doi. org/10.1111/1467-9884.00314
16. Glickman, M.E.: Parameter estimation in large dynamic paired comparison experiments. J. Royal Stat. Soc. Ser. C (Appl. Stat.) **48**(3), 377–394 (2002). https:// doi.org/10.1111/1467-9876.00159
17. Goddard, J.: Regression models for forecasting goals and match results in association football. Int. J. Forecast. **21**(2), 331–340 (2005). https://doi.org/10.1016/j. ijforecast.2004.08.002
18. Goddard, J., Asimakopoulos, I.: Forecasting football results and the efficiency of fixed-odds betting. J. Forecast. **23**(1), 51–66 (2004). https://doi.org/10.1002/for. 877
19. Govan, A.Y., Langville, A.N., Meyer, C.D.: Offense-defense approach to ranking team sports. J. Q. Anal. Sports **5**(1) (2009). https://doi.org/10.2202/1559-0410. 1151
20. Graham, I., Stott, H.: Predicting bookmaker odds and efficiency for UK football. Appl. Econ. **40**(1), 99–109 (2008). https://doi.org/10.1080/00036840701728799
21. Groll, A., Ley, C., Schauberger, G., Van Eetvelde, H.: Prediction of the FIFA World Cup 2018 - a random forest approach with an emphasis on estimated team ability parameters. arXiv:1806.03208 [stat], June 2018
22. Herbrich, R., Minka, T., Graepel, T.: TrueSkill^{TM} : a Bayesian skill rating system. In: Schölkopf, B., Platt, J.C., Hoffman, T. (eds.) Advances in Neural Information Processing Systems, vol. 19, pp. 569–576. MIT Press (2007)
23. Hubáček, O., Šourek, G., Železný, F.: Learning to predict soccer results from relational data with gradient boosted trees. Mach. Learn. **108**(1), 29–47 (2019). https://doi.org/10.1007/s10994-018-5704-6
24. Hvattum, L.M., Arntzen, H.: Using ELO ratings for match result prediction in association football. Int. J. Forecast. **26**(3), 460–470 (2010). https://doi.org/10. 1016/j.ijforecast.2009.10.002
25. Joy, B., Weil, E., Giulianotti, R.C., Alegi, P.C., Rollin, J.: Football. https://www. britannica.com/sports/football-soccer
26. Karlis, D., Ntzoufras, I.: Analysis of sports data by using bivariate Poisson models. J. Royal Stat. Soc. **52**(3), 381–393 (2003). https://doi.org/10.1111/1467-9884. 00366
27. Keener, J.: The Perron–Frobenius theorem and the ranking of football teams. SIAM Rev. **35**(1), 80–93 (1993). https://doi.org/10.1137/1035004
28. Kuypers, T.: Information and efficiency: an empirical study of a fixed odds betting market. Appl. Econ. **32**(11), 1353–1363 (2000). https://doi.org/10.1080/ 00036840050151449
29. Langville, A.N., Meyer, C.D.: Who's #1?: The Science of Rating and Ranking. Princeton University Press, Princeton (2012)
30. Lasek, J., Szlávik, Z., Bhulai, S.: The predictive power of ranking systems in association football. Int. J. Appl. Pattern Recogn. **1**(1), 27–46 (2013). https://doi.org/ 10.1504/IJAPR.2013.052339
31. Lee, A.J.: Modeling scores in the premier league: is Manchester united really the best? Chance **10**(1), 15–19 (1997). https://doi.org/10.1080/09332480.1997. 10554791
32. Leitner, C., Zeileis, A., Hornik, K.: Forecasting sports tournaments by ratings of (prob)abilities: a comparison for the EURO 2008. Int. J. Forecast. **26**(3), 471–481 (2010). https://doi.org/10.1016/j.ijforecast.2009.10.001

33. Ley, C., Van de Wiele, T., Van Eetvelde, H.: Ranking soccer teams on basis of their current strength: a comparison of maximum likelihood approaches. eprint arXiv:1705.09575, May 2017
34. Maher, M.J.: Modelling association football scores. Stat. Neerl. **36**(3), 109–118 (1982). https://doi.org/10.1111/j.1467-9574.1982.tb00782.x
35. McCullagh, P.: Regression models for ordinal data. J. Royal Stat. Soc. **42**(2), 109–142 (1980)
36. Park, J., Newman, M.E.J.: A network-based ranking system for American college football. J. Stat. Mech. Theory Exp. **2005**(10), P10014–P10014 (2005). https://doi.org/10.1088/1742-5468/2005/10/P10014
37. Pope, P.F., Peel, D.A.: Information, prices and efficiency in a fixed-odds betting market. Economica **56**(223), 323–341 (1989). https://doi.org/10.2307/2554281
38. Rue, H., Salvesen, O.: Prediction and retrospective analysis of soccer matches in a league. J. Royal Stat. Soc. Ser. D (Stat.) **49**(3), 399–418 (2000). https://doi.org/10.1111/1467-9884.00243
39. Spann, M., Skiera, B.: Sports forecasting: a comparison of the forecast accuracy of prediction markets, betting odds and tipsters. J. Forecast. **28**(1), 55–72 (2008). https://doi.org/10.1002/for.1091
40. Stefani, R.T.: Improved least squares football, basketball, and soccer predictions. IEEE Trans. Syst. Man Cybern. **10**(2), 116–123 (1980). https://doi.org/10.1109/TSMC.1980.4308442
41. Tsokos, A., Narayanan, S., Kosmidis, G.I.B., Cucuringu, M., Whitaker, G., Kiraly, F.: Modeling outcomes of soccer matches. Mach. Learn. **108**(1), 77–95 (2019)
42. UBS AG: and the winner is.... investing in emerging markets (special edition, 2018 World Cup in Russia), May 2018
43. Van Haaren, J., Davis, J.: Predicting the final league tables of domestic football leagues. In: Proceedings of the 5th International Conference on Mathematics in Sport, pp. 202–207 (2015)
44. Zeileis, A., Leitner, C., Hornik, K.: Probabilistic forecasts for the 2018 FIFA World Cup based on the bookmaker consensus model, p. 19
45. Zyga, L.: New algorithm ranks sports teams like Google's PageRank, December 2009. https://phys.org/news/2009-12-algorithm-sports-teams-google-pagerank.html

Distinguishing Between Roles of Football Players in Play-by-Play Match Event Data

Bart Aalbers[(✉)] and Jan Van Haaren[(✉)]

SciSports, Hengelosestraat 500, 7521 AN Enschede, Netherlands
{b.aalbers,j.vanhaaren}@scisports.com

Abstract. Over the last few decades, the player recruitment process in professional football has evolved into a multi-billion industry and has thus become of vital importance. To gain insights into the general level of their candidate reinforcements, many professional football clubs have access to extensive video footage and advanced statistics. However, the question whether a given player would fit the team's playing style often still remains unanswered. In this paper, we aim to bridge that gap by proposing a set of 21 player roles and introducing a method for automatically identifying the most applicable roles for each player from play-by-play event data collected during matches.

Keywords: Football analytics · Player recruitment · Player roles

1 Introduction

Over the last few decades, professional football has turned into a multi-billion industry. During the 2017/2018 season, the total revenue of all twenty English Premier League clubs combined exceeded a record high of 5.3 billion euro with 1.8 billion euro spent on the acquisition of players in the 2017 summer transfer window alone [2,10]. Hence, the player recruitment process has become of vital importance to professional football clubs. That is, football clubs who fail to strengthen their squads with the right players put their futures at stake.

Compared to twenty years ago, scouts at football clubs have plenty more tools at their disposal to find appropriate players for their clubs. Scouts nowadays can watch extensive video footage for a large number of players via platforms like Wyscout[1] and obtain access to advanced statistics, performance metrics and player rankings via platforms like SciSports Insight.[2] Although these tools provide many insights into the abilities and general level of a player, they largely fail to answer the question whether the player would fit a certain team's playing style. For example, star players like Cristiano Ronaldo and Lionel Messi operate at a comparable level but clearly act and behave in different ways on the pitch.

---

[1] https://www.wyscout.com.
[2] https://insight.scisports.com.

© Springer Nature Switzerland AG 2019
U. Brefeld et al. (Eds.): MLSA 2018, LNAI 11330, pp. 31–41, 2019.
https://doi.org/10.1007/978-3-030-17274-9_3

To help answer this question, we compose a set of 21 candidate player roles in consultation with the SciSports Datascouting department and introduce an approach to automatically identify the most applicable roles for each player from play-by-play event data collected during matches. To optimally leverage the available domain knowledge, we adopt a supervised learning approach and pose this task as a probabilistic classification task. For each player, we first extract a set of candidate features from the match event data and then perform a separate classification task for each of the candidate player roles.

2   Dataset

Our dataset consists of play-by-play match event data for the 2017/2018 season of the English Premier League, Spanish LaLiga, German 1. Bundesliga, Italian Serie A, French Ligue Un, Dutch Eredivisie and Belgian Pro League provided by SciSports' partner Wyscout. Our dataset represents each match as a sequence of roughly 1500 on-the-ball events such as shots, passes, crosses, tackles, and interceptions. For each event, our dataset contains a reference to the team, a reference to the player, the type of the event (e.g., a pass), the subtype of the event (e.g., a cross), 59 boolean indicators providing additional context (e.g., accurate or inaccurate and left foot or right foot), the start location on the pitch and when relevant also the end location on the pitch.

3   Player Roles

We compile a list of 21 player roles based on an extensive sports-media search (e.g., [1,5,12]) and the Football Manager 2018 football game, which is often renowned for its realism and used by football clubs in their player recruitment process (e.g., [6,11]). In consultation with the football experts from the SciSports Datascouting department, we refine the obtained list and also introduce two additional roles, which are the mobile striker and the ball-playing goalkeeper. The mobile striker is dynamic, quick and aims to exploit the space behind the defensive line. The ball-playing goalkeeper has excellent ball control and passing abilities and often functions as an outlet for defenders under pressure.

For each of these roles, we define a set of key characteristics that can be derived from play-by-play match event data. Most of the defined player roles describe players who excel in technical ability, game intelligence, strength, pace or endurance. Figure 1 presents an overview of the 21 player roles, where related roles are connected by arrows and roles for the same position are in the same color. For example, a full back and wing back have similar defensive duties, which are marking the opposite wingers, recovering the ball, and preventing crosses and through balls. In contrast, a wing back and winger have similar offensive duties, which are providing width to the team, attempting to dribble past the opposite backs, and creating chances by crossing the ball into the opposite box.

Due to space constraints, we focus this paper on the five roles for central midfielders. We now present each of these five roles in further detail.

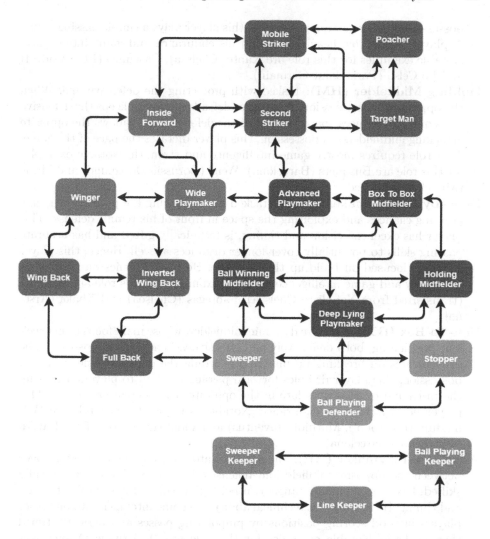

Fig. 1. A graphical overview of the 21 proposed player roles for six different positions. Roles that have tasks and duties in common are connected by arrows. Roles that apply to the same position are in the same color. Roles for goalkeepers are in green, roles for defenders are in light blue, roles for backs are in red, roles for central midfielders are in dark blue, roles for wingers are in pink and roles for forwards are in purple. (Color figure online)

Ball-Winning Midfielder (BWM): mainly tasked with gaining possession. When the opponent has possession, this player is actively and aggressively defending by closing down opponents and cutting off pass supply lines. This player is a master in disturbing the build-up of the opposing team, occasionally making strategic fouls such that his team can reorganize. When in possession or after gaining possession, this player plays a simple passing game. A player in this role heavily relies on his endurance and game intelligence. Notable examples for this role are Kanté (Chelsea), Casemiro (Real Madrid) and Lo Celso (Paris Saint-Germain).

Holding Midfielder (HM): tasked with protecting the defensive line. When the opponent has possession, this player defends passively, keeps the defensive line compact, reduces space in front of the defence and shadows the opposite attacking midfielders. In possession, this player dictates the pace of the game. This role requires mostly game intelligence and strength. Notable examples for this role are Busquets (Barcelona), Weigl (Borussia Dortmund) and Matić (Manchester United).

Deep-Lying Playmaker (DLP): tasked with dictating the pace of the game, creating chances and exploiting the space in front of his team's defense. This player has excellent vision and timing, is technically gifted and has accurate passing skills to potentially cover longer distances as well. Hence, this player is rather focused on build-up than defense. He heavily relies on his technical ability and game intelligence. Notable examples for this role are Jorginho (transferred from Napoli to Chelsea), Fàbregas (Chelsea) and Xhaka (Arsenal).

Box-To-Box (BTB): a more dynamic midfielder, whose main focus is on excellent positioning, both defensively and offensively. When not in possession, he concentrates on breaking up play and guarding the defensive line. When in possession, this player dribbles forward passing the ball to players higher up the pitch and often arrives late in the opposite box to create a chance. This player heavily relies on endurance. Notable examples for this role are Wijnaldum (Liverpool), Matuidi (Juventus) and Vidal (transferred from Bayern München to Barcelona).

Advanced Playmaker (AP): the prime creator of the team, occupying space between the opposite midfield and defensive line. This player is technically skilled, has a good passing range, can hold up a ball and has excellent vision and timing. He relies on his technical ability and game intelligence to put other players in good scoring positions by pinpointing passes and perfectly timed through balls. Notable examples for this role are De Bruyne (Manchester City), Luis Alberto (Lazio Roma) and Coutinho (Barcelona).

4   Approach

In this section, we propose a method to automatically derive the most applicable player roles for each player from play-by-play match event data. We adopt a supervised learning approach to optimally leverage the available domain knowledge within the SciSports Datascouting department. More specifically, we address this task by performing a probabilistic classification task for each of the candidate roles introduced in Sect. 3. In the remainder of this section, we explain the feature engineering and probabilistic classification stages in more detail.

4.1   Feature Engineering

The goal of the feature engineering stage is to obtain a set of features that both characterizes the candidate player roles and distinguishes between them. The most important challenges are to distinguish between the qualities and quantities of a player's actions and to account for the strength of a player's teammates and opponents as well as the tactics and strategies employed by a player's team.

We perform the feature engineering stage in two steps. In the first step, we compute a set of basic statistics and more advanced performance metrics. We base these statistics and metrics on the specific tasks players in the different roles perform. We normalize these basic statistics and metrics in two different ways to improve their comparability across players. In the second step, we standardize the obtained statistics and metrics to facilitate the learning process and construct features by combining them based on domain knowledge.

Computing Statistics and Metrics. We compute a set of 242 basic statistics and metrics for each player across a set of matches (e.g., one full season of matches). This set includes statistics like the number of saves from close range, the number of duels on the own half, the number of passes given into the opposite box, the number of long balls received in the final third, the number of offensive ground duels on the flank and the number of high-quality attempts. For example, a ball-winning midfielder is tasked with regaining possession, aggressively pressing opponents and laying off simple passes to more creative teammates. Hence, we compute statistics like the number of interceptions in the possession zone (i.e., the middle of the pitch), the number of defensive duels won in the possession zone, and the number of low-risk passes from the possession zone.

We normalize these statistics and metrics in two different ways to obtain 484 features. First, we normalize the values by the total number of actions performed by the player. This normalization helps to improve the comparability across players on strong and weak teams, where strong teams tend to perform more on-the-ball actions. Second, we normalize the values by the total number of actions performed by the player's team. This normalization helps to identify key players that are often in possession of the ball (e.g., playmakers).

In addition, we compute 31 metrics including centrality measures in the passing network, defensive presence in the team and a player's wide contribution.

Constructing Features. We construct a set of 181 key features by combining the statistics and metrics. In order to facilitate the learning process, we standardize our set of 515 features. More specifically, we project the feature values to the $[\mu - 2\sigma, \mu + 2\sigma]$ interval, where $\mu = 0$ and $\sigma = 1$. We project any extreme values that fall outside the interval to the interval bounds.

We use the resulting set of standardized features to construct the key feature set. For example, we combine the statistics representing the number of attempts from the area immediately in front of goal and the expected-goals metrics for these attempts into a key feature reflecting the frequency of high-quality attempts from short range.

4.2   Probabilistic Classification

We obtain labels for a portion of our dataset from the SciSports Datascouting department. In particular, we request the Datascouting department to assign the primary role to each of the players in our dataset they feel confident enough about to assess. Although football players often fulfill several different roles during the course of a season or even within a single match, we only collect each player's primary role to simplify our learning setting.

We perform a separate binary classification task for each of the 21 roles. For each role, we construct a separate dataset in two steps. First, we obtain all examples of players that were assigned that role and label them as positive examples. Second, we obtain all examples of players that were assigned one of the "disconnected" roles in Fig. 1 as well as 25% of the examples for players that were assigned one of the "connected" roles and label them as negative examples.

We notice that keeping 25% of the examples for players having a connected role improves the accuracy of our classification models. If we keep all connected examples, the learner focuses on the specific tasks that distinguish between the role of interest and the connected roles. If we remove all connected examples, the learner focuses on distinguishing between positions rather than specific roles.

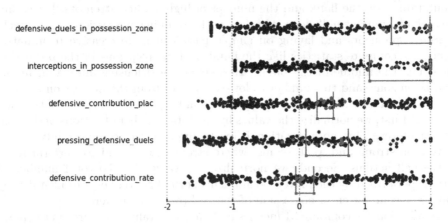

Fig. 2. The five most relevant features for ball-winning midfielders. The blue dots represent players labeled as ball-winning midfielders, while the black dots represent players labeled as any of the other roles. The blue lines represent the optimal region. (Color figure online)

We perform a final transformation of the key feature values before they are fed into the classification algorithm. For each key feature and each role, we first define the optimal feature value range and then compute the distance between the actual feature value and the optimal range. We determine the optimal range by computing percentile bounds on the labeled training data. We set a parameter β, which translates into a lower bound of β and an upper bound of $1 - \beta$, as can

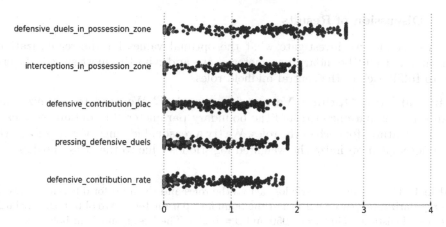

Fig. 3. The five most relevant features for ball-winning midfielders after the transformation. The blue dots represent players labeled as ball-winning midfielders, while the black dots represent players labeled as any of the other roles. The optimal value is 0. (Color figure online)

be seen in Fig. 2. For example, $\beta = 0.25$ translates into the $[0.25, 0.75]$ range. We set the feature values that fall within the optimal range to 0 and all other feature values to their distance to the closest bound of the optimal range. Hence, we project the original $[-2, 2]$ range to a $[0, 4]$ range, where 0 is the optimal value, as can be seen in Fig. 3.

5   Experimental Evaluation

This section presents an experimental evaluation of our approach. We present the methodology, formulate the research questions and discuss the results.

5.1   Methodology

We construct one example for each player in our dataset. After dropping all players who played fewer than 900 min, we obtain a set of 1910 examples. The SciSports Datascouting department manually labeled 356 of these examples (i.e., 18.6% of the examples) as one of the 21 roles introduced in Sect. 3.

We prefer algorithms learning interpretable models to facilitate the explanation of the inner workings of our approach to football coaches and scouts. More specifically, we train a Stochastic Gradient Descent classifier using logistic loss as the objective function for each role. Hence, the coefficients of the model reflect the importances of the features.

For each role, we perform ten-fold cross-validation and use the Synthetic Minority Over-Sampling Technique (SMOTE) to correct the balance between the positive and negative examples [4]. After performing the transformation explained in Sect. 4.2, we obtain a set of over 300 examples.

5.2   Discussion of Results

In this section, we investigate what the optimal values for the regularization parameter α and boundary parameter β are and who the most-suited players are to fulfill each of the central midfield roles.

What are the Optimal Values for α and β? We need to optimize the regularization parameter α and the boundary parameter β to obtain the most-relevant features for each of the roles. We try a reasonable range of values for both parameters and optimize the weighted logistic loss due to the class imbalance.

Table 1. The weighted logistic loss across the roles and the folds for different values of the regularization parameter α and the boundary parameter β. We obtain the optimal weighted logistic loss for $\alpha = 0.050$ and $\beta = 0.250$. The best result is in bold.

α	β	Weighted logistic loss
1.000	0.250	0.2469
0.500	0.250	0.1800
0.100	0.250	0.0977
0.050	**0.250**	**0.0831**
0.010	0.250	0.0835
0.005	0.250	0.1034
0.001	0.250	0.2199
0.050	0.100	0.0865
0.050	0.200	0.0833
0.050	**0.250**	**0.0831**
0.050	0.300	0.0840
0.050	0.350	0.0863
0.050	0.400	0.0890

Table 1 shows the weighted logistic loss across the ten folds and 21 roles for several different values for the regularization parameter α and the boundary parameter β. The top half of the table shows that 0.050 is the optimal value for the regularization parameter α, while the bottom half of the table shows that 0.250 is the optimal value for the boundary parameter β.

What Are the Most Suitable Players for the Central Midfielder Roles?
Table 2 shows the predicted probabilities of fulfilling any of the five central midfielder roles for 20 players in our dataset. For each role, the table shows the top-ranked labeled player as well as the top-three-ranked unlabeled players.

The table shows that most players fulfill different types of roles during the course of a season. In general, players who rate high on one role also score high on at least one other role. For example, top-ranked box-to-box midfielders like Jonas

Table 2. The probabilities for the five central midfielder roles for 20 players in our dataset. For each role, we show the top-ranked labeled player as well as the top-three-ranked unlabeled players. The highest probability for each player is in bold.

Player	Labeled	BWM	HM	DLP	BTB	AP
Trigueros	Yes	**0.98**	0.85	0.87	0.89	0.29
Sergi Darder	No	**0.98**	0.90	0.82	0.94	0.11
Guilherme	No	**0.97**	0.94	0.32	0.93	0.03
E. Skhiri	No	**0.97**	0.93	0.38	0.96	0.04
N. Matic	Yes	0.91	**0.98**	0.82	0.42	0.01
J. Guilavogui	No	0.96	**0.97**	0.71	0.76	0.05
Rodrigo	No	0.90	**0.96**	0.23	0.48	0.00
G. Pizarro	No	0.90	**0.96**	0.24	0.41	0.00
João Moutinho	Yes	0.91	0.88	**0.98**	0.41	0.14
J. Henderson	No	0.68	0.79	**0.97**	0.13	0.07
S. Kums	No	0.87	0.90	**0.96**	0.37	0.03
D. Demme	No	0.73	0.76	**0.96**	0.22	0.04
J. Martin	Yes	0.84	0.48	0.40	**0.98**	0.21
Zurutuza	No	0.90	0.75	0.39	**0.98**	0.23
T. Rincón	No	0.93	0.75	0.29	**0.98**	0.11
T. Bakayoko	No	0.90	0.57	0.29	**0.98**	0.20
C. Eriksen	Yes	0.08	0.01	0.43	0.18	**0.98**
J. Pastore	No	0.12	0.02	0.56	0.31	**0.97**
H. Vanaken	No	0.31	0.03	0.58	0.35	**0.95**
V. Birsa	No	0.18	0.01	0.21	0.57	**0.94**

Martin (Strasbourg), David Zurutuza (Real Sociedad), Tomás Rincón (Torino) and Tiémoué Bakayoko (Chelsea) also score high for the ball-winning midfielder role. The advanced playmaker (AP) role seems to be an exception though. The players who rate high on this role do not rate high on any of the other roles.

Furthermore, the table also shows that our approach is capable of handling players playing in different leagues. With deep-lying playmaker Sven Kums of Anderlecht and advanced playmaker Hans Vanaken of Club Brugge, two players from the smaller Belgian Pro League appear among the top-ranked players for their respective roles. The table lists six players from the Spanish LaLiga, four players from the English Premier League, three players from the French Ligue Un, three players from the Italian Serie A, two players from the German 1. Bundesliga, two players from the Belgian Pro League, and no players from the Dutch Eredivisie.

6   Related Work

The task of automatically deriving roles of players during football matches has largely remained unexplored to date. To the best of our knowledge, our proposed approach is the first attempt to identify specific roles of players during matches fully automatic in a data-driven fashion using a supervised learning approach.

In contrast, most of the work in this area focuses on automatically deriving positions and formations from data in an unsupervised setting. Bialkowski et al. [3] present an approach for automatically detecting formations from spatio-temporal tracking data collected during football matches. Pappalardo et al. [8] present a similar approach for deriving player positions from play-by-play event data collected during football matches. Our approach goes beyond their approaches by not only deriving each player's standard position in a formation (e.g., a left winger in a 4-3-3 formation) but also his specific role within that position (e.g., a holding midfielder or a ball-winning midfielder).

Furthermore, researchers have compared football players in terms of their playing styles. Mazurek [7] investigates which player is most similar to Lionel Messi in terms of 24 advanced match statistics. Peña and Navarro [9] analyze passing motifs to find an appropriate replacement for Xavi who exhibits a similar playing style. In addition, several articles discussing playing styles and player roles have recently appeared in the mainstream football media (e.g., [1,5,12]).

7   Conclusion

This paper proposed 21 player roles to characterize the playing styles of football players and presented an approach to automatically derive the most applicable roles for each player from play-by-play event data. Our supervised learning approach proceeds in two steps. First, we compute a large set of statistics and metrics that both characterize the different roles and help distinguish between the roles from match data. Second, we perform a binary classification task for each role leveraging labeled examples obtained from the SciSports Datascouting department. Our approach goes beyond existing techniques by not only deriving each player's standard position (e.g., an attacking midfielder in a 4-2-3-1 formation) but also his specific role within that position (e.g., an advanced playmaker). Our experimental evaluation demonstrates our approach for deriving five roles for central midfielders from data collected during the 2017/2018 season.

In the future, we plan to further improve and extend our approach. We will investigate different learning algorithms to tackle the classification task (e.g., XGBoost) as well as different learning settings. More specifically, we aim to obtain a richer set of labels from the SciSports Datascouting department. In addition to each player's primary role, we also wish to collect possible alternative player roles, which would turn our task into a multi-label classification setting.

References

1. Arsenal.com: Jack and Xhaka: Strengths and Styles (2018). https://www.arsenal.com/news/news-archive/20161125/jack-and-xhaka-strengths-and-styles. Accessed 3 Aug 2018
2. Barnard, M., Dwyer, M., Wilson, J., Winn, C.: Annual Review of Football Finance 2018 (2018). https://www2.deloitte.com/content/dam/Deloitte/uk/Documents/sports-business-group/deloitte-uk-sbg-annual-review-of-football-finance-2018.PDF. Accessed 3 Aug 2018
3. Bialkowski, A., Lucey, P., Carr, P., Yue, Y., Matthews, I.: Win at home and draw away: automatic formation analysis highlighting the differences in home and away team behaviors. In: Proceedings of 8th Annual MIT Sloan Sports Analytics Conference. Citeseer (2014)
4. Chawla, N.V., Bowyer, K.W., Hall, L.O., Kegelmeyer, W.P.: SMOTE: synthetic minority over-sampling technique. JAIR (J. Artif. Intell. Res.) **16**, 321–357 (2002)
5. Cox, M.: World Cup Favourites Choosing Defensive-Minded Midfielders over Deep-Lying Playmakers (2018). http://www.espn.co.uk/football/fifa-world-cup/4/blog/post/3520743. Accessed 3 Aug 2018
6. Goal.com: How Lionel Messi and Andres Iniesta Could Have Ended up in Glasgow (2012). http://www.goal.com/en-gb/news/3277/la-liga/2012/10/23/3471081/how-lionel-messi-and-andres-iniesta-could-have-ended-up-in
7. Mazurek, J.: Which football player bears most resemblance to Messi? A statistical analysis. arXiv preprint arXiv:1802.00967 (2018)
8. Pappalardo, L., Cintia, P., Ferragina, P., Pedreschi, D., Giannotti, F.: PlayeRank: multi-dimensional and role-aware rating of soccer player performance. arXiv preprint arXiv:1802.04987 (2018)
9. Peña, J.L., Navarro, R.S.: Who can replace Xavi? A passing motif analysis of football players. arXiv preprint arXiv:1506.07768 (2015)
10. Poli, R., Besson, R., Ravenel, L.: The CIES Football Observatory 2017/18 season (2018). http://www.football-observatory.com/IMG/pdf/cies_football_analytics_2018.pdf. Accessed 3 Aug 2018
11. Sullivan, J.: Beautiful and Mathematical: Football as a Numbers Game (2016). https://www.bbc.com/news/science-environment-37327939
12. Thompson, M.: Four Types of Forward in the Champions League Final (2018). https://www.footballwhispers.com/blog/four-types-of-forward-in-the-champions-league-final. Accessed 3 Aug 2018

Player Valuation in European Football

Edward Nsolo, Patrick Lambrix$^{(\boxtimes)}$, and Niklas Carlsson

Linköping University, Linköping, Sweden
patrick.lambrix@liu.se

Abstract. As the success of a team depends on the performance of individual players, the valuation of player performance has become an important research topic. In this paper, we compare and contrast which attributes and skills best predict the success of individual players in their positions in five European top football leagues. Further, we evaluate different machine learning algorithms regarding prediction performance. Our results highlight features distinguishing top-tier players and show that prediction performance is higher for forwards than for other positions, suggesting that equally good prediction of defensive players may require more advanced metrics.

Keywords: Sports analytics · Data mining · Player valuation

1 Introduction

The success of a football team depends a lot on the individual players making up that team. However, not all positions on a team are the same and there are significant differences in the style of game being played in different leagues. It is therefore important to take into account both the position and the league when evaluating individual players.

In this paper, we compare and contrast which attributes and skills best predict the success of individual players, within and across leagues and positions. First, we investigate which performance features are important in five European top leagues (Italy, Spain, England, Germany, and France) for defenders, midfielders, forwards, and goal keepers, respectively. Second, as part of our analysis, we evaluate different techniques for generating prediction models (based on different machine learning algorithms) for players belonging to the top segment of players. To capture further differentiation among the best players of a league, in our experiments, we have investigated top sets corresponding to 10%, 25% and 50% of the most highly ranked players in each considered category.

Our results provide interesting insights into what features may distinguish a top-tier player (in the top 10%, for example), from a good player (in the top 25%), or from an average player, within and across the leagues. The results also suggest that predicting performance may be easier for forwards than for other positions. Our work distinguishes itself from other work on player valuation or player performance, by working with tiers of players (in contrast to individual

© Springer Nature Switzerland AG 2019
U. Brefeld et al. (Eds.): MLSA 2018, LNAI 11330, pp. 42–54, 2019.
https://doi.org/10.1007/978-3-030-17274-9_4

ratings) as well as by dealing with many skills (in contrast to some work with focus on a particular skill).

The remainder of the paper is organized as follows. Section 2 presents related work. Section 3 discusses the data sets and the data preparation. Sections 4 and 5 present the feature selection and prediction methods, respectively, and show and discuss the corresponding results. Finally, conclusions are presented in Sect. 6.

2   Related Work

In many sports, work has started on valuation of player performance. For the sake of brevity, we address the related work in football.

Much work focuses on game play, including rating game actions [9,28], pass prediction [5,7,13,18,31], shot prediction [25], expectation for goals given a game state [8], or more general game strategies [1,3,4,11,12,14,15,24]. Other works try to predict the outcome of a game by estimating the probability of scoring for the individual goal scoring opportunities [10], by relating games to the players in the teams [26], or by using team rating systems [21].

Regarding player performance, in [29] a weighted ± measure for player performance is introduced. The skill of field vision is investigated in [20] while in [32] the authors develop a method to, based on the current skill set, determine the skill set of a player in the future. The performance versus market value for forwards is discussed in [17]. Further, the influence of the heterogeneity of a team on team performance is discussed in [19].

3   Data Collection and Preparation

3.1   Data Collection

Data was collected from five top European leagues, i.e., the highest English (English Premier League, EPL), Spanish (La Liga), German (BundesLiga), Italian (Serie A) and French (Ligue 1) leagues for the 2015–16 season. WhoScored (https://www.whoscored.com) was used as the main data provider. Most of its data is acquired from Opta (http://www.optasports.com) which is used by many secondary data providers. WhoScored and other secondary data providers use internal schemes developed by a group of soccer experts to rate player and team performances. The differences in rating players is relatively small. Therefore, we used the WhoScored rating as the gold standard in our experiments. The data contained information about 2,606 players. The list of attributes is given in Table 1. These are the attributes of WhoScored except for some where we aggregated the attributes, as for instance, goals represents goals by right foot, goals by left foot, goals by head and goals by other body parts.

We generated six groups of data sets, one group for each league as well as one for all leagues together. This allows us to find differences and commonalities between the different leagues. For each group, we divided the players with respect to their position on the field into 222 goalkeepers (GKs), 970 defenders (DFs),

Table 1. List of attributes. The defensive, offensive and passing categories are as they are defined in WhoScored. Note that some attributes are in several of these categories. We added the two other categories for the remaining attributes used in WhoScored.

Type	Attributes
Identifying	Player id, nationality, player name, age, position, height, weight, team name, league
Defensive	Tackles, interceptions, offsides won, clearance, blocks, own goals, dribbles, fouls committed
Offensive	Goals, assists, shots per game, key passes, fouled, offsides committed, dispossessed, bad control, dribbles
Passing	Assists, key passes, passes per game, pass success rate, crosses, long balls, through balls
All	Man of the match, fulltime, halftime, minutes played, aerials won, red cards, yellow cards
Class	Rating

1,109 midfielders (MDs), and 305 forwards (FWs). For the field players, players whose primary role and position on the pitch were defending and back were categorized as defenders. Players whose primary role and position on the pitch were playmaking and central were categorized as midfielders. Finally, players whose primary role and position on the pitch were attacking and front, were categorized as forwards.

3.2 Data Preparation

For each data set representing a position (GK, DF, MD, FW) within a group (specific league or all leagues), we generated three final data sets representing the top 10%, 25% and 50% players for that position and group. We used Weka [16] with TigerJython in this phase.

We filled in some missing values in the data, such as nationality for some players, and used a uniform way to represent missing values over the data sets. We detected duplicates of players which transferred to other teams during the season and merged the data for these players. This affected 3 GKs, 20 DFs, 20 MDs, and 71 FWs. Each data set was normalized using the min-max method which transformed given numerical attribute values to a range of 0 to 1. The normalized value of value x is (x - min)/(max - min), where min and max are the minimum and maximum value for the attribute, respectively. The rating of WhoScored was used for the binary class attribute (deciding whether the player is in the top X% or not). The data was discretized to reflect this. We used SMOTE [6] to overcome the class imbalance for the 10% and 25% data sets. It is an oversampling technique that synthetically determines copies of the instances of the minority class to be added to the data set to match the quantity of instances of the majority class. We refer to these final data sets (after all preparation steps) as LEAGUE-POSITION-Xpc where LEAGUE has the values EPL, Bundesliga, La-Liga, Ligue-1, Serie-A and All; POSITION has the values GK, DF, MD, and FW; and X has the values 10, 25, 50 (e.g., EPL-GK-25pc).

4   Feature Selection

For each data set we used two ways implemented in Weka [16], a filter method and a wrapper method, to select possible features that are important for player performance.

4.1   Filter Method

With the filter method we computed absolute values of the Pearson correlation coefficients between each attribute and the class attribute. In our application we retained all attributes that had a Pearson correlation coefficient of at least 0.3.

The results for the combined leagues are shown in Table 2. For the results regarding the individual leagues, we refer to [27].

This method is fast and runs in milliseconds for the different cases.

4.2   Wrapper Method

Wrapper methods use machine learning (ML) algorithms and evaluate the performance of the algorithms on the data set using different subsets of attributes. We used the Weka setting where we started from the empty set and used best-first search with backtracking after five consecutive non-improving nodes in the search tree. We used seven ML algorithms that can handle numeric, categorical, and binary attributes and that were chosen from different types; i.e., BayesNet (Bayesian nets), NaiveBayes (Naive Bayes classifier), Logistic (builds linear logistic regression models), IBk (k-nearest-neighbor), J48 (a C4.5 decision tree learner), Part (rules from partial decision trees), and RandomForest (random forests). The merit of a subset was evaluated using WrapperSubsetEval that uses the classifier with, in our case, five-fold cross-validation. Then we computed a support count for each attribute reflecting how often it occurred in a selected attribute subset computed by the different ML algorithms. Attributes that were selected at least twice were retained as important attributes.

The results for the data sets for the combined leagues are shown in Table 3. For the results for the individual leagues we refer to [27].

The run times for this method depend on the ML algorithms and the data sets ranging from below a minute to several hours (see [27]). Logistic and RandomForest were the slowest. For the Bayesian approaches there was not much difference between the different data sets.

Table 2. Attributes filter method for combined leagues. Attributes in *italics* are in common with the other data sets for the same position (All-X-10pc, All-X-25pc, All-X-50pc). Attributes in **bold** are in common with the same data set for the wrapper method.

Selected attributes	Data set
minutes played, fulltime, clearance, dribble, yellow cards	All-GK-10pc
\<none>	All-GK-25pc
man of the match	All-GK-50pc
interceptions, ***man of the match***, ***aerial won***, *crosses*, *tackles*, halftime, assists, *fulltime*, minutes played	All-DF-10pc
man of the match, *crosses*, aerial won, *interceptions*, **minutes played**, *fulltime*, *tackles*, **goals**, **halftime**, **shots per game**	All-DF-25pc
crosses, **fulltime**, minutes played, *interceptions*, *man of the match*, aerial won, *tackles*, **goals**, **through balls**	All-DF-50pc
man of the match, key passes, *fulltime*, *minutes played*, *shots per game*, assists, *crosses*, *goals*, **fouled**, *dispossessed*, halftime, *long balls*, *through balls*, *bad control*, tackles	All-MD-10pc
fulltime, *minutes played*, ***man of the match***, **key passes**, *crosses*, **assists**, *fouled*, **shots per game**, *goals*, *dispossessed*, tackles, **halftime**, *through balls*, *long balls*, *bad control*, dribble	All-MD-25pc
fulltime, *minutes played*, *crosses*, **key passes**, **assists**, *man of the match*, *fouled*, **tackles**, **shots per game**, interceptions, yellow cards, *goals*, *through balls*, dribble, *dispossessed*, *bad control*, *long balls*, fouls committed, clearance	All-MD-50pc
crosses, *shots per game*, *fulltime*, *goals*, ***man of the match***, *minutes played*, *assists*, *passes per game*, *key passes*, *fouled*, *aerial won*, *through balls*, *dispossessed*, *bad control*, *offsides committed*, *clearance*, *yellow cards*, halftime, *age*, *tackles*, *fouls committed*	All-FW-10pc
fulltime, *minutes played*, **shots per game**, *crosses*, **goals**, *passes per game*, *key passes*, *man of the match*, *assists*, *dispossessed*, *bad control*, *fouled*, **offsides committed**, *aerial won*, *through balls*, *tackles*, *clearance*, *yellow cards*, *fouls committed*, **age**	All-All-FW-25pc
crosses, *fulltime*, *minutes played*, **shots per game**, *key passes*, *passes per game*, **bad control**, *goals*, *dispossessed*, *fouled*, *aerial won*, **assists**, *offsides committed*, *through balls*, **fouls committed**, *man of the match*, *clearance*, *tackles*, *yellow cards*, *age*	All-FW-50pc

Table 3. Attributes wrapper method for combined leagues. Attributes in *italics* are in common with the other data sets for the same position (All-X-10pc, All-X-25pc, All-X-50pc). Attributes in **bold** are in common with the same data set for the filter method. The numbers in parentheses show the support counts. For attributes without a number the support count is 2.

Selected attributes	Data set
player name (5), player id (3), tackles, own goals, halftime, long balls	All-GK-10pc
player name (4), team name (3), player id (3) own goals, through balls	All-GK-25pc
man of the match (7), *player name* (3), aerial won, player id	All-GK-50pc
interceptions (4), player name (4), **aerial won** (3), *man of the match* (3), *shots per game* (3), *tackles* (3), *team name* (3), age, passes per game, blocks, red cards, yellow cards	All-DF-10pc
player name (5), ***man of the match*** (5), ***shots per game*** (4), ***tackles*** (4), *team name* (4), **crosses** (3), **halftime** (3), **goals**, own goals, dispossessions, **fulltime**, offsides committed, *interceptions*, **minutes played**	All-DF-25pc
man of the match (7), *interceptions* (7), **crosses** (6), *shots per game* (5), ***tackles*** (4), assists (4), blocks (3), **goals**, *team name*, **through balls**, offsides committed, league	All-DF-50pc
player name (5), team name, **fouled**, own goals, *man of the match*	All-MD-10pc
man of the match (6), player name (4), **halftime** (4), **assists** (3), **shots per game**, crosses, pass success rate, league, red cards, **goals**, blocks, **key passes**	All-MD-25pc
fulltime (7), **crosses** (6), *man of the match* (4), **key passes** (3), **tackles** (3), **shots per game** (3), pass success rate, **assists**	All-MD-50pc
man of the match (5), player name (5), own goals (4), blocks (3), player id	All-FW-10pc
man of the match (7), player id (4), name (4), aerial won (3), **shots per game** (3), halftime (3), **goals**, **offsides committed**, age	All-FW-25pc
man of the match (6), **crosses** (5), **shots per game** (3), **assists** (3), interceptions (3), red cards, offsides won, **fouls committed**, weight, **bad control**	All-FW-50pc

4.3   Discussion

Filter Method: As expected, the selected attributes did not include any of the identifying attributes such as team name, nationality, player name and league, which all received low correlation coefficients.

As we used the absolute value of the Pearson correlation coefficients, the impact of the selected attributes could be positive as well as negative on the class variable. In our results most of these are considered positive (e.g., goals for forwards). We note that for the offensive and passing attributes the values are the highest for the FWs, while for the defensive attributes they are usually highest for the DFs.

The top 10 of selected attributes over all the different leagues [27] (i.e., they are selected for most combinations of league, position and top X%) contain attributes related to all player responsibilities, such as tackles for defense, shots per game and goals for offense, crosses for passing, and key passes and assists for both offense and passing. Further, there are other attributes such as man of the match, minutes played, full time and aerials won. Identifying attributes as well as red cards, own goals, and offsides won are rarely selected.

For all leagues, fewer attributes are selected for GKs and DFs than for MDs and FWs. In Tables 2 and 3 we have marked the attributes that are in common for the same position in *italics*. In particular, for FWs and MDs many attributes are selected for each of the top 10%, top 25% and top 50% data sets, for DFs there are fewer common attributes while very few for GKs.

Interception is selected more often for Serie A than for other leagues and is important for all positions in Serie A. Through balls is selected less in La Liga than in other leagues. Height is only selected for top 10% GKs, DFs and FWs in the Bundesliga and top 10% GKs in the EPL. Tackles are selected for all DFs in the Bundesliga, Ligue 1 and Serie A, while only for some of the DF data sets for EPL and La Liga. Offsides committed is selected for all FWs in all leagues, except for the EPL where it is only selected for the top 50% data set.

Wrapper Method: The subsets produced by WrapperSubsetEval had merit scores of over 68% for the data sets of the combined leagues and 77% for the data sets of the individual leagues. IBk, NaiveBayes and Logistic had on average the highest merit scores (of over 90%) across all data sets, except for some GK and top 50% data sets. The selected subsets for the top-10% data sets for all four categories of players for all individual and for the combined leagues had the highest merit scores while the selected subsets for the top-50% data sets had the lowest merit scores. This might be caused by the amount of instances added when handling class imbalances. Further, the merit scores for the selected subsets for FWs were always higher than those for MDs, which in their turn were higher than for DFs and GKs.

In general, few selected attributes are common for the data sets related to the same position. For instance, for the Bundesliga there were no selected attributes in common for the different data sets at all. For the EPL, clearance was in common for GKs, aerials won and man of the match for DFs and man of the match for MDs. In contrast to the filter method, identifying attributes do appear in the selected lists.

Both: We note that the selected attributes for the combined leagues and the individual leagues are different. This suggests that players doing well in one league will not necessarily do well in another leagues and may reflect the fact that the playing style is different in different leagues.

A larger ratio of the selected attributes for the wrapper methods are in common with the selected attributes for the filter methods for the same data set. For the combined leagues the overlap is larger than for the individual leagues and the overlap is largest for DFs and MDs. For the individual leagues the Bundesliga has a small overlap which is mostly situated in the MD data sets.

5   Prediction

5.1   Methods

For each LEAGUE-POSITION-Xpc data set, based on the results of the feature selection procedure we created two data sets with the selected attributes from the wrapper and filter methods, respectively. Using Weka [16] we ran several ML algorithms: RandomForest, BayesNet, Logistic, DecisionTable (a decision table majority classifier), IBk, KStar (nearest neighbor with generalized distance), NaiveBayes, J48, Part, and ZeroR (predicts the majority class for nominals or the average value for numerics). ZeroR was used as a baseline. We split each data set into a training (66% of the instances) and a testing set (34% of the instances). All experiments were run 10 times and we calculated averages for the performance values.

5.2   Results

As performance metrics we used the standard measures of accuracy, precision, recall, F1 score (or f-measure), and AUC-ROC. Given the true and false positives (TP and FP) and the true and false negatives (TN and FN), the accuracy is $(TP + TN)/(TP + TN + FP + FN)$, the precision is $TP/(TP + FP)$ and the recall is $TP/(TP + FN)$. The F1 score is a harmonic mean over precision and recall. AUC-ROC plots recall against precision.

Figures 1 and 2 show summary statistics for F1 scores and prediction performance, respectively, from full factor experiments, in which we evaluated every combination of (i) league or the combined leagues, (ii) position, (iii) top-set category, (iv) filter/wrapper selection, and (v) ML technique being applied. In total, this resulted in $6 \times 4 \times 3 \times 2 \times 10 = 1,4400$ prediction evaluations (14,400 when taking into account 10 runs per configuration). Other and more detailed results are presented in an extended version [27].

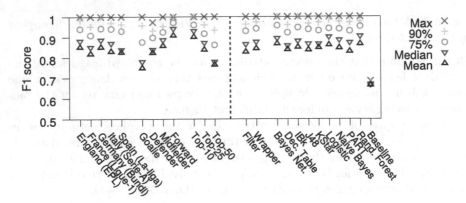

Fig. 1. Distribution statistics when keeping the reported factor fixed and varying all other factors.

5.3   Discussion

Figure 1 shows the distribution statistics for the individual leagues when keeping the reported factor fixed and varying all other factors. In particular, for each such factor and level, we present the maximum F1 score, the 90%-ile F1 score, the 75%-ile F1 score, the median F1 score (equal to the 50%-ile score), and the average F1 score. Here, the first three factors (i.e., the left hand side of Fig. 1) allow us to compare and contrast the prediction scores obtained for different (i) leagues, (ii) player positions, and (iii) top-set categories. These results suggest a clear ordering in the prediction scores based on position and top set. For example, FWs have the highest F1 scores, MDs the second highest, DFs the second lowest, and GKs the lowest F1 scores. This suggests that ML may be more successfully applied to evaluate more offensive positions, but may be less successful to predict the skills of more defensive positions (with GKs the other extreme of the spectrum). The techniques also appear much better at predicting players in the top-10 set than predicting players in the top-50 set (again with a clear ordering of the three sets). Smaller differences are observed across the leagues, although the EPL and Bundesliga appear to be the two leagues for which it is easiest to use basic ML to predict top players.

The right hand side of Fig. 1 compares the prediction success of the different methods. In general, the wrapper method is more successful than the filter method, and BayesNet and RandomForest provide the highest F1 scores across the distribution metrics. All methods significantly outperform the baseline.

Finally, we look closer at how well the above methods (the baseline excluded) can predict the top X players of each position in each of the leagues. Fig. 2 presents these results. Here, we show results for the best predictor results (max) for each top-X set, where X = 10%, 25%, and 50%, and the corresponding medians (when considering all filter/wrapper combinations with the different ML techniques). Again, the top-10% set is much easier to predict across the leagues (with medians above 0.9 across all leagues and positions, except GKs in France).

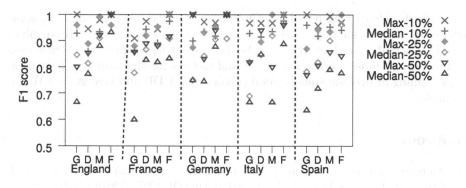

Fig. 2. F1 scores of the top-X player sets (X = 10, 25, 50%) of each position and league.

As the top sets become larger, the F1 scores decrease, and again the more offensive positions obtain substantially higher prediction scores across the leagues.

6   Conclusion

In this paper we generated attributes and skills sets that best predict the success of individual players in their positions in five European top leagues. In contrast to other work we focused on the top tiers of players (top 10%, 25% and 50%). Further, we evaluated different ML algorithms on their performance of predicting the tiers using the generated attributes. For the sake of brevity we have shown some of the results in the paper while for more results we refer to [27]. Among other things, our prediction results show (i) a clear ordering in the prediction scores based on position (e.g., F1 of FW > F1 of MD > F1 of DF > F1 of GK) and top set (e.g., F1 of top 10% > F1 of top 25% > F1 of top 50%), (ii) that basic ML techniques are most successful to predict top players in EPL and Bundesliga (although good performance across the leagues), (iii) that the wrapper method is more successful than the filter method, and (iv) that BayesNet and RandomForest provide the highest F1 scores of the considered ML techniques.

One limitation of the approach is that the method is based on a ranking of experts, which is the current state of expertise. However, as in the case of baseball, the opinions of which properties constitute a 'good' player may change [22]. Future work could consider longitudinal drift in the rankings. It is also interesting to investigate the correlation of the rankings with the players' market value. Another limitation is that the approach does not take the quality of the team mates into account (although *team name* was sometimes a selected attribute). Although important, we are not aware of work that takes this explicitly into account. However, there is work on related questions such as team performance [5,7,19,21,26] and, in other sports, the performance of pairs or triples of players (e.g., [23,30] for ice hockey and [2] for basketball). Another direction for future work is to apply the methodology only on game-related attributes. Some of the identifying attributes may not be that interesting for teams for identifying possible future players for the team. Further, for some attributes, such as shots per game, we may investigate using normalized values based on actual play time. It could also be interesting to look at different tiers and to develop more advanced features and metrics for the more defensive positions, aiming to provide equally good prediction for DF and GK as for FW, for example.

References

1. Andrienko, G., et al.: Visual analysis of pressure in football. Data Min. Knowl. Discov. **31**(6), 1793–1839 (2017). https://doi.org/10.1007/s10618-017-0513-2
2. Ayer, R.: Big 2's and Big 3's: analyzing how a team's best players complement each other. In: MIT Sloan Sports Analytics Conference (2012)
3. Bialkowski, A., Lucey, P., Carr, P., Yue, Y., Sridharan, S., Matthews, I.: Large-scale analysis of soccer matches using spatiotemporal tracking data. In: Kumar, R., Toivonen, H., Pei, J., Huang, J.Z., Wu, X. (eds.) Proceedings of the 2014 IEEE International Conference on Data Mining, pp. 725–730 (2014)
4. Bojinov, I., Bornn, L.: The pressing game: optimal defensive disruption in soccer. In: 10th MIT Sloan Sports Analytics Conference (2016)
5. Brandt, M., Brefeld, U.: Graph-based approaches for analyzing team interaction on the example of soccer. In: Davis, J., van Haaren, J., Zimmermann, A. (eds.) Proceedings of the 2nd Workshop on Machine Learning and Data Mining for Sports Analytics. CEUR Workshop Proceedings, vol. 1970, pp. 10–17 (2015)
6. Chawla, N.V., Bowyer, K.W., Hall, L.O., Kegelmeyer, W.P.: SMOTE: synthetic minority over-sampling technique. J Artif Intell. Res. **16**, 321–357 (2002). https://doi.org/10.1613/jair.953
7. Cintia, P., Rinzivillo, S., Pappalardo, L.: Network-based measures for predicting the outcomes of football games. In: Davis, J., van Haaren, J., Zimmermann, A. (eds.) Proceedings of the 2nd Workshop on Machine Learning and Data Mining for Sports Analytics. CEUR Workshop Proceedings, vol. 1970, pp. 46–54 (2015)
8. Decroos, T., Dzyuba, V., Van Haaren, J., Davis, J.: Predicting soccer highlights from spatio-temporal match event streams. In: Proceedings of the 31st AAAI Conference on Artificial Intelligence, pp. 1302–1308 (2017)
9. Decroos, T., Van Haaren, J., Dzyuba, V., Davis, J.: STARSS: a spatio-temporal action rating system for soccer. In: Davis, J., Kaytoue, M., Zimmermann, A. (eds.) Proceedings of the 4th Workshop on Machine Learning and Data Mining for Sports Analytics. CEUR Workshop Proceedings, vol. 1971, pp. 11–20 (2017)
10. Eggels, H., van Elk, R., Pechenizkiy, M.: Explaining soccer match outcomes with goal scoring opportunities predictive analytics. In: van Haaren, J., Kaytoue, M., Davis, J. (eds.) Proceedings of the 3rd Workshop on Machine Learning and Data Mining for Sports Analytics. CEUR Workshop Proceedings, vol. 1842 (2016)
11. Fernando, T., Wei, X., Fookes, C., Sridharan, S., Lucey, P.: Discovering methods of scoring in soccer using tracking data. In: Lucey, P., Yue, Y., Wiens, J., Morgan, S. (eds.) Proceedings of the 2nd KDD Workshop on Large Scale Sports Analytics (2015)
12. Gyarmati, L., Anguera, X.: Automatic extraction of the passing strategies of soccer teams. In: Lucey, P., Yue, Y., Wiens, J., Morgan, S. (eds.) Proceedings of the 2nd KDD Workshop on Large Scale Sports Analytics (2015)

13. Gyarmati, L., Stanojevic, R.: QPass: a merit-based evaluation of soccer passes. In: Lucey, P., Yue, Y., Wiens, J., Morgan, S. (eds.) Proceedings of the 3rd KDD Workshop on Large Scale Sports Analytics (2016)

14. Haaren, J.V., Davis, J., Hannosset, S.: Strategy discovery in professional soccer match data. In: Lucey, P., Yue, Y., Wiens, J., Morgan, S. (eds.) Proceedings of the 3rd KDD Workshop on Large Scale Sports Analytics (2016)

15. Van Haaren, J., Dzyuba, V., Hannosset, S., Davis, J.: Automatically discovering offensive patterns in soccer match data. In: Fromont, E., De Bie, T., van Leeuwen, M. (eds.) IDA 2015. LNCS, vol. 9385, pp. 286–297. Springer, Cham (2015). https://doi.org/10.1007/978-3-319-24465-5_25

16. Hall, M., Frank, E., Holmes, G., Pfahringer, B., Reutemann, P., Witten, I.H.: The WEKA data mining software: an update. SIGKDD Explor. 11(1), 10–18 (2009). https://doi.org/10.1145/1656274.1656278

17. He, M., Cachucho, R., Knobbe, A.: Football player's performance and market value. In: Davis, J., van Haaren, J., Zimmermann, A. (eds.) Proceedings of the 2nd Workshop on Machine Learning and Data Mining for Sports Analytics. CEUR Workshop Proceedings, vol. 1970, pp. 87–95 (2015)

18. Horton, M., Gudmundsson, J., Chawla, S., Estephan, J.: Classification of passes in football matches using spatiotemporal data. In: Lucey, P., Yue, Y., Wiens, J., Morgan, S. (eds.) Proceedings of the 1st KDD Workshop on Large Scale Sports Analytics (2014)

19. Ingersoll, K., Malesky, E., Saiegh, S.M.: Heterogeneity and team performance: evaluating the effect of cultural diversity in the world's top soccer league. J. Sports Anal. 3(2), 67–92 (2017). https://doi.org/10.3233/JSA-170052

20. Jordet, G., Bloomfield, J., Heijmerikx, J.: The hidden foundation of field vision in English Premier League (EPL) soccer players. In: 7th MIT Sloan Sports Analytics Conference (2013)

21. Lasek, J.: EURO 2016 predictions using team rating systems. In: van Haaren, J., Kaytoue, M., Davis, J. (eds.) Proceedings of the 3rd Workshop on Machine Learning and Data Mining for Sports Analytics. CEUR Workshop Proceedings, vol. 1842 (2016)

22. Lewis, M.: Moneyball: The Art of Winning an Unfair Game. W.W. Norton & Company, New York (2003). ISBN 978-0-393-05765-2

23. Ljung, D., Carlsson, N., Lambrix, P.: Player pairs valuation in ice hockey. In: Brefeld, U., Davis, J., van Haaren, J., Zimmermann, A. (eds.) Proceedings of the 5th Workshop on Machine Learning and Data Mining for Sports Analytics. LNAI, vol. 11330, pp. 82–92. Springer, Cham (2018)

24. Lucey, P., Bialkowski, A., Carr, P., Foote, E., Matthews, I.: Characterizing multi-agent team behavior from partial team tracings: evidence from the English Premier League. In: 26th AAAI Conference on Artificial Intelligence, pp. 1387–1393 (2012)

25. Lucey, P., Bialkowski, A., Monfort, M., Carr, P., Matthews, I.: Quality vs quantity: improved shot prediction in soccer using strategic features from spatiotemporal data. In: 9th MIT Sloan Sports Analytics Conference (2015)

26. Maystre, L., Kristof, V., Ferrer, A.J.G., Grossglauser, M.: The player kernel: learning team strengths based on implicit player contributions. In: van Haaren, J., Kaytoue, M., Davis, J. (eds.) Proceedings of the 3rd Workshop on Machine Learning and Data Mining for Sports Analytics. CEUR Workshop Proceedings, vol. 1842 (2016)

27. Nsolo, E., Lambrix, P., Carlsson, N.: Player valuation in European football (extended version) (2018). www.ida.liu.se/research/sportsanalytics/projects/conferences/MLSA18-soccer

28. Sarkar, S., Chakraborty, S.: Pitch actions that distinguish high scoring teams: findings from five European football leagues in 2015–16. J. Sports Anal. 4(1), 1–14 (2018). https://doi.org/10.3233/JSA-16161
29. Schultze, S.R., Wellbrock, C.M.: A weighted plus/minus metric for individual soccer player performance. J. Sports Anal. 4(2), 121–131 (2018). https://doi.org/10.3233/JSA-170225
30. Thomas, A., Ventura, S.L., Jensen, S., Ma, S.: Competing process hazard function models for player ratings in ice hockey. Ann. Appl. Stat. 7(3), 1497–1524 (2013)
31. Vercruyssen, V., Raedt, L.D., Davis, J.: Qualitative spatial reasoning for soccer pass prediction. In: van Haaren, J., Kaytoue, M., Davis, J. (eds.) Proceedings of the 3rd Workshop on Machine Learning and Data Mining for Sports Analytics. CEUR Workshop Proceedings, vol. 1842 (2016)
32. Vroonen, R., Decroos, T., Haaren, J.V., Davis, J.: Predicting the potential of professional soccer players. In: Davis, J., Kaytoue, M., Zimmermann, A. (eds.) Proceedings of the 4th Workshop on Machine Learning and Data Mining for Sports Analytics. CEUR Workshop Proceedings, vol. 1971, pp. 1–10 (2017)

Ranking the Teams in European Football Leagues with Agony

Stefan Neumann[1]([✉]), Julian Ritter[2], and Kailash Budhathoki[3]

[1] Faculty of Computer Science, University of Vienna, Vienna, Austria
`stefan.neumann@univie.ac.at`
[2] Laboratoire d'informatique, École polytechnique, Palaiseau, France
`julian.ritter@polytechnique.edu`
[3] Max Planck Institute for Informatics and Saarland University,
Saarbrücken, Germany
`kbudhath@mpi-inf.mpg.de`

Abstract. Ranking football (soccer) teams is usually done using league tables as provided by the organizers of the league. These league tables are designed to yield a total ordering of the teams in order to assign unique ranks to the teams. Hence, the number of levels in the league table equals the number of teams. However, in sports analytics one would be interested in categorizing the teams into a small number of *substantially* different levels of *playing quality*. In this paper, our goal is to solve this issue for European football leagues. Our approach is based on a generalized version of agony which was introduced by Gupte et al. (WWW'11). Our experiments yield natural rankings of the teams in four major European football leagues into a small number of interpretable quality levels.

Keywords: Football · Leagues · Ranking · Hierarchy · Agony

1 Introduction

When thinking about football (soccer) leagues, one of the first things that comes to mind is the league table. The position of a team in the league table determines the fate of managers, the happiness of fans and how much money the club can spend in the next season. Since the league table must determine which teams win the league, which qualify for European competitions and which are relegated, it contains a lot of tie breakers so that each team is assigned a unique rank.

However, if we wish to assess the performances of teams from a data analytics point of view, the league table has multiples downsides: First, we would like to evaluate the number of *substantially* different levels of playing quality in the league. The league table offers as many levels as there are teams in the league, but one would expect a much smaller number of quality levels (for example, the top clubs who compete for the trophy, the clubs fighting against relegation and the ones somewhere in between). Second, we would like to understand against which kind of opposition teams perform well. For instance, some teams might

U. Brefeld et al. (Eds.): MLSA 2018, LNAI 11330, pp. 55–66, 2019.
https://doi.org/10.1007/978-3-030-17274-9_5

earn a lot of points by very consistently beating bad teams but at the same time losing most games against the very good teams. The league table would cover this up since the number of points awarded against good/bad opposition are exactly the same.

1.1 Our Contributions

In this paper, we take a step towards resolving the aforementioned issues. That is, we show how to find a small number of interpretable levels of playing quality in the top 4 major European leagues. We also point out how some of our results indicate that certain teams perform well (or badly) against very good teams.

To obtain our results, we define graphs based on the match results in the football leagues we consider. We then apply algorithms for finding hierarchies in graphs. These algorithms will provide us with a ranking of the teams into a small number of levels which are highly interpretable.

In Sect. 3 we briefly discuss for how many matchdays such rankings can be obtained without any *conflicting results*. For example, a conflict refers to results with team A beating team B, team B beating team C and team C beating team A. For such conflicting results it is inherently difficult to find good rankings. We will formalize this in Sect. 3.

We address this issue with conflicts by introducing a generalized version of the agony problem [7] in Sect. 4.

To be a bit more specific, we will be looking at two types of graphs: (1) Directed graphs, where the vertices are the teams of the league and where we insert an edge from team A to team B if A beats B, and (2) undirected graphs, where we insert edges if A and B played a draw. The graph algorithms will then partition the vertices into levels such that (1) most directed edges point from "good" teams (at high levels) towards "bad" teams (at lower levels), and (2) most undirected edges are between clubs on the same level.

Section 5 presents our results. We provide hierarchies of the teams in four major European football leagues in the 2017/2018 season.

2 Related Work

Ranking Football Teams. Several approaches for assessing the quality of football teams have been proposed.

Hvattum and Arntzen [11] and Leitner et al. [13] considered a modification of the ELO rating system. ELO was initially developed for chess [5] and [11,13] modified it for usage in the football domain. Some implementations of ELO ratings are also available online [1,2].

Constantinou and Fenton [4] introduced pi-ratings and they demonstrated that pi-ratings perform better than the ELO rankings of [11,13] when predicting the outcomes of future matches. Pi-ratings have also been used in state of the art methods for predicting the outcomes of football matches [3,10].

ELO and pi-ratings assign quality scores to teams and these scores can be used to assess the strength of the teams. However, unlike the results presented in this paper, these scores are continuous-valued and do not group the teams into a small number of substantially different playing levels.

Hierarchies in Graphs. Finding hierarchies in directed graphs is an old and fundamental problem. In this line of work, one is given a directed graph $G = (V, E)$ and one wants to assign levels to the vertices in V such that certain properties are satisfied. A drawing of the graph respecting these levels is called a *hierarchy*.

For example, if G is a directed acyclic graph (DAG), we can draw G such that all of its edges point downwards. Then the drawing immediately provides the levels of the vertices in V. However, such an approach fails as soon as G contains a cycle. See Sect. 3 for more details.

One classic way to obtain hierarchies from graphs which contain cycles is to solve the *feedback arc set problem*. This problem asks to find a subgraph $H = (V, E')$ of G such that H is a DAG and the number of *non-DAG edges* $E \setminus E'$ is minimized. Unfortunately, this problem is NP-hard to solve [12] and known polynomial time algorithms only achieve poly-logarithmic approximations [6]. Hanauer [8,9] conducted an experimental study on algorithms for the feedback arc set problem.

Gupte et al. [7] introduced *agony* which can be viewed as a tractable version of the feedback arc set problem. Agony does not minimize the number of non-DAG edges (i.e., the edges pointing upwards) but it weights each of these non-DAG edges by the number levels it points upwards; we provide a technical definition of agony in Sect. 4. [7] showed that agony can be used to identify hierarchies in social networks and to find hierarchies in American football leagues; they also provided a polynomial time algorithm. Later, Tatti [15–17] provided faster and more general algorithms for computing agony.

However, neither the feedback arc set problem nor agony as studied in [7, 15–17] allow for finding hierarchies in which certain vertices in the graph are supposed to be on the same level. Thus, while these algorithms can be used for sport leagues *without* draws (such as American football or basketball), they do not apply to sports like football where draws frequently occur.

3 A Partial Order for a Partial Season

In this section, we discuss for how many matchdays we can rank the teams of a football league based on the results of their games *without contradicting results*.

To start out, let us first formalize what we mean by contradicting results. Consider the set of teams T and define a binary relation $>$ on the set of teams T as follows. For teams $A, B \in T$, let $A > B$ if A won against B. Now, we have *contradicting results* if there are teams $A_1, \ldots, A_k$ such that $A_1 > A_2 > \cdots > A_k > A_1$. For example, such a contradiction occurs if Chelsea beat Liverpool, who in turn beat Manchester United, who in turn beat Chelsea.

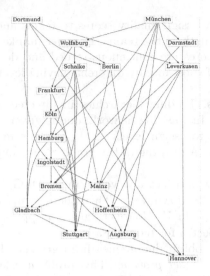

Fig. 1. The first seven matchdays of the German Bundesliga season 2015/16.

Note that if the results contain a contradiction, we cannot rank the teams in the league properly. In the previous example, it is not clear how we should rank Chelsea, Liverpool and Manchester United because either of them could be placed on level 1.

Hence, for the rest of this section, let us only consider the first ℓ matchdays of a league where the results contain no contradicting results. In such a case the result graph G is a directed acyclic graph (DAG). More formally, $G = (T, E)$ is the graph with the teams in T as vertices and containing an edge from team A to team B if $A > B$.

If G is a DAG, then it is quite simple to obtain a ranking and a drawing for it. We can place the teams without incoming edges at level 1 (i.e., those teams which did not lose). Next, at level 2, we place all vertices which only have incoming edges from vertices already placed. Then we continue in this fashion for all levels $i \geq 3$. Note that all edges must be pointing downwards.

To obtain interesting rankings for a given league, we consider the largest number L such that the results of the first L match days have no contradicting results. Figure 1 presents the result graph G for the German Bundesliga season 2015/16. We picked this season for the reason that it had $L = 7$, yielding an elaborate diagram of the first seven matchdays. Subsequent Bundesliga seasons each had $L = 4$.

Unfortunately, contradicting results are quite common in football leagues or tournaments. Skinner and Freeman [14, Sect. 3] study this problem for the FIFA World Cup results from 1938–2006 and in case of random game outcomes.

4   Agony for Sport Leagues with Draws

In this section, we show how we can obtain meaningful hierarchies even in the presence of contradicting results as per Sect. 3. To do this, we will consider a generalized version of agony which we will be using throughout the rest of the paper. Agony was first studied by Gupte et al. [7] (see Sect. 2).

Definition. Let $G = (V, E_{\mathrm{dir}}, w)$ be a directed graph with edge weights $w(u, v)$ and let $H = (V, E_{\mathrm{undir}})$ be an *undirected* unweighted graph. Both graphs are defined over the same set of vertices. The ranking that we will compute is related to G and H in the following way. Edges from u to v in the directed graph G indicate that u should be above v (this corresponds to team u winning against team v). The edges in the undirected graph H, on the other hand, indicate that u and v should be on the same level in the hierarchy (this corresponds to teams u and v drawing). We will now make this more formal.

A *ranking* r is a function $r : V \to \mathbb{N}$ which maps a vertex v to a level $r(v)$. In the following, when we draw the vertices in V according to some ranking function r, we will start with vertices in level 1 at the top, followed by the vertices on level 2 below that, and so on. Thus, an edge from u at level 5 to a node v on level 2 is pointing *upwards*; vice versa, an edge from u at level 1 to v at level 9 is pointing *downwards*.

Now consider a fixed ranking function r. We will define two different quantities: First, the agony caused by r on G and, second, the agony of r on H. Intuitively, edges pointing upwards w.r.t. r will cause agony and we will want to find a ranking function which has few of such edges.

The *agony of r on G* is defined as

$$A_{\mathrm{dir}}(G, r) = \sum_{(u,v) \in E_{\mathrm{dir}}} \max\{r(u) - r(v) + 1, 0\} \cdot w(u, v). \tag{1}$$

To develop a better understanding of the sum in Eq. 1, consider an edge directed from u to v and suppose that all weights $w(u, v)$ are 1. Now, if u has a higher rank than v (e.g., $r(u) = 5$ and $r(v) = 1$), then the edge is pointing upwards in the hierarchy. Intuitively, this should introduce some agony and, indeed, we have $r(u) - r(v) + 1 > 0$. On the other hand, if u has a smaller rank than v (e.g., $r(u) = 1$, $r(v) = 2$), the quantity $r(u) - r(v) + 1$ is non-positive and no agony will be caused by this edge. This aligns well with our intuition since the edge is pointing downwards in the drawing of the vertices. Finally, if two vertices are on the same level, their agony will be exactly 1 due to the $+1$ term. Note that in Eq. 1 we also multiply this level difference with the corresponding edge weight.

We note that $A_{\mathrm{dir}}(G, r)$ is exactly the definition of agony used in [7] and generalized to weighted graphs.

The *agony of r on H* is given by

$$A_{\mathrm{undir}}(H, r) = \sum_{(u,v) \in E_{\mathrm{undir}}} |r(u) - r(v)|. \tag{2}$$

Note that $A_{\mathrm{undir}}(G, r)$ does not introduce any agony for edges (u, v) which have both endpoints on the same level (because then $|r(u) - r(v)| = 0$). However, for edges (u, v) with endpoints on different levels, it will count the level difference.

Now our goal is to find a ranking function r^* which minimizes the sum of the agony on both graphs, i.e., we look for the minimizer of

$$A(G, H) = \min_{r \in V^{\mathbb{N}}} A_{\mathrm{dir}}(G, r) + A_{\mathrm{undir}}(H, r). \tag{3}$$

We note that when setting the undirected graph H to an empty graph (i.e., $E_{\mathrm{undir}} = \emptyset$), then Eq. 3 becomes exactly the same as in [7].

Integer Linear Program (ILP). Based on the definitions of agony, we now state the ILP that we solve to compute the agony $A(G, H)$. Solving the ILP also provides the ranking function r^* which is minimizing the agony.

The ILP is stated in Eq. 4 and it is using the following variables: $r(u)$ is the level assigned to vertex u; $x(u, v)$ is the agony caused by the directed edge $(u, v) \in E_{\mathrm{dir}}$; $y(u, v)$ corresponds to the agony caused by the undirected edge $(u, v) \in E_{\mathrm{undir}}$.

$$
\begin{aligned}
\text{minimize} \quad & \sum_{(u,v) \in E_{\mathrm{dir}}} x(u, v) + \sum_{(u,v) \in E_{\mathrm{undir}}} y(u, v) & & (4) \\
\text{subject to} \quad & x(u, v) \geq (r(u) - r(v) + 1) \cdot w(u, v) & & \text{for all } (u, v) \in E_{\mathrm{dir}} \\
& x(u, v) \geq 0 & & \text{for all } (u, v) \in E_{\mathrm{dir}} \\
& y(u, v) \geq r(u) - r(v) & & \text{for all } (u, v) \in E_{\mathrm{undir}} \\
& y(u, v) \geq r(v) - r(u) & & \text{for all } (u, v) \in E_{\mathrm{undir}} \\
& r(u) \geq 1 & & \text{for all } u \in V \\
& r(u), x(u, v), y(u, v) \in \mathbb{Z}
\end{aligned}
$$

We now show that the ILP in Eq. 4 correctly computes the agony.

Lemma 1. *The ILP stated in Eq. 4 correctly computes the agony as defined in Eq. 3.*

Proof (Sketch). Let $r(u)$, $x(u, v)$, $y(u, v)$ be a solution for the ILP.

First, we show that $x(u, v) = \max\{r(u) - r(v) + 1, 0\} \cdot w(u, v)$ for each $(u, v) \in E_{\mathrm{dir}}$. Indeed, the constraints imply that $x(u, v) \geq 0$ and $x(u, v) \geq (r(u) - r(v) + 1) \cdot w(u, v)$. Since $x(u, v)$ is not present in any other constraint and since we are minimizing the objective function, we obtain the desired equality. (The same constraints were also derived in [7].)

Second, we show that $y(u, v) = |r(u) - r(v)|$ for each edge $(u, v) \in E_{\mathrm{undir}}$. Note that either both $r(u) - r(v)$ and $r(v) - r(u)$ are zero or exactly one of them is positive. This implies that $y(u, v)$ is non-negative and also that $y(u, v) \geq |r(u) - r(v)|$. As before, the equality follows since $y(u, v)$ does not occur in any other constraint and because we are minimizing in the objective function.

It follows that the ILP and agony compute exactly the same. $\square$

4.1   Agony for Football Leagues

Next we discuss how an algorithm for the agony problem from the previous subsection can be used to create the ranking for a football league.

Based on the results of the football league, we create two graphs G and H and each of these graphs has one vertex for each team. The graph G is directed with edge weights $w(u, v)$ and the graph H is undirected and unweighted.

First, assume that all teams play each other exactly once. Suppose team A beats team B with a goal difference of z. For example, Borussia Dortmund beats Schalke with 5:0. In this case, we insert an edge from team A to team B with weight $z = 5$ into the directed graph G. If team A and team B play draw, we insert an undirected edge into the undirected graph H. Thus, G will correspond to games with a winner and H will correspond to games that ended in draws.

Now, to handle multiple games between the same teams, we just add up the goal differences of all of their games. For example, if Borussia Dortmund and Schalke play three games ending 5:1, 0:2, and 4:4, then the goal difference on the edge will be 2. I.e., we will consider the "direct comparison" of the teams.

5   Evaluation

We implemented our algorithm in Python, and used the PuLP library for solving the ILP. The ILP can be solved in a few seconds because the problem instances are small.

To obtain better and more robust rankings, we set the weights in the directed graph G slightly differently than discussed above. If the goal difference of teams u and v is d, then we set the weight of the edge (u, v) to $w(u, v) = 1 + \log(d)$, where the logarithm has base 2.

There were three reasons for this choice of $w(u, v)$: (1) When we ran the experiments with weight function $w(u, v) = d$, the ranking was not very robust. For example, if a team lost only a *single* game with a relatively large goal difference this often lead to the team loosing one level in the hierarchy. Hence, a single bad performance could affect the ranking of a team excessively much. (2) The choice of $w(u, v) = 1 + \log(d)$ has several nice properties. For narrow wins/losses with goal difference $d = 1, 2$, $w(u, v) = d$. Thus, for narrow games the choice of $w(u, v)$ has no influence on the ranking. For higher wins/losses, however, $d \geq 3$ and $w(u, v) < d$; thus, larger losses are penalized less and the rankings become more robst than discussed in point (1). (3) We also tried out different weight functions such as $w(u, v) = d^{\beta}$ for different parameters $\beta \in (0, 5]$ but the function we used gave the most natural rankings.

5.1   Case Study on European Football Leagues

We consider four major European football leagues, namely Premier League (England), La Liga (Spain), Serie A (Italy), and Bundesliga (Germany) for evaluating the ILP. To this end, we used the data from www.football-data.co.uk. The

Rank	Teams
1	Bayern Munich
2	Dortmund, Hoffenheim, Schalke 04, Leverkusen, Leipzig
3	Mainz, Hannover, Wolfsburg, M'gladbach, Augsburg, Hertha, Stuttgart, Ein Frankfurt, Werder Bremen
4	Hamburg, Freiburg, FC Köln

Team	P	W	D	L	GD	Pts
Bay. Munich	34	27	3	4	64	84
Schalke 04	34	18	9	7	16	63
Hoffenheim	34	15	10	9	18	55
Dortmund	34	15	10	9	17	55
Leverkusen	34	15	10	9	14	55
Leipzig	34	15	8	11	4	53
Stuttgart	34	15	6	13	0	51
Ein Frankfurt	34	14	7	13	0	49
M'gladbach	34	13	8	13	-5	47
Hertha	34	10	13	11	-3	43
Werd Bremen	34	10	12	12	-3	42
Augsburg	34	10	11	13	-3	41
Hannover	34	10	9	15	-10	39
Mainz	34	9	9	16	-14	36
Freiburg	34	8	12	14	-24	36
Wolfsburg	34	6	15	13	-12	33
Hamburg	34	8	7	19	-24	31
FC Köln	34	5	7	22	-35	32

Rank	Teams
1	Manchester City, Manchester United
2	Liverpool, Arsenal, Tottenham
3	Leicester, Newcastle, West Ham, Burnley, Bournemouth, Chelsea, Crystal Palace, Watford, Everton
4	Southampton, Stoke, Huddersfield, West Brom, Swansea, Brighton

Team	P	W	D	L	GD	Pts
Man City	38	32	4	2	79	100
Man United	38	25	6	7	40	81
Tottenham	38	23	8	7	38	77
Liverpool	38	21	12	5	46	75
Chelsea	38	21	7	10	24	70
Arsenal	38	19	6	13	23	63
Burnley	38	14	12	12	-3	54
Everton	38	13	10	15	-14	49
Leicester	38	12	11	15	-4	47
Newcastle	38	12	8	18	-8	44
Crystal Palace	38	11	11	16	-10	44
Bournemouth	38	11	11	16	-16	44
West Ham	38	10	12	16	-20	42
Watford	38	11	8	19	-20	41
Brighton	38	9	13	16	-20	40
Huddersfield	38	9	10	19	-30	37
Southampton	38	7	15	16	-19	36
Swansea	38	8	9	21	-28	33
Stoke	38	7	12	19	-33	33
West Brom	38	6	13	19	-25	31

Fig. 2. Results on (top) Bundesliga, and (bottom) English Premier League. Accompanying the results on the right hand side are the corresponding league tables for the year 2017/2018.

dataset for a football league contains statistics such as match stats, full time results, and total goals from all the games played in that league.

The results of the ILP on all four leagues are shown in Figs. 2 and 3. For comparison, we also provide the league tables for the season 2017/2018 alongside the resulting hierarchy produced by the ILP.

Bundesliga. In Bundesliga, the ILP discovers four levels. Bayern Munich is the only team in the top level (LEVEL 1). This is also corroborated by the massive points difference (Pts) as well as the goal difference (GD) between Bayern Munich and the second team in Bundesliga. The teams from the second till the sixth position in the league table are in the second level (LEVEL 2). Given that they have similar statistics (the number of wins, draws, losses, and the goal difference), the result of ILP is in agreement with our intuition. LEVEL 4 consists of the two bottom-last teams Koln and Hamburg, as well as Freiburg (rank 15 in the table). Freiburg is in the fourth level since they lost the direct comparison against five teams on LEVEL 3 (some of them with large margins). The remaining midfield teams are in the third level (LEVEL 3).

English Premier League (EPL). The ILP yields an interesting result on the EPL. Although second-place Manchester United has much less points than the premier league winner Manchester City, they are both in the first level. This may seem counter-intuitive at first. There's more to it, however. Manchester United lost the home game, but won the away game against Manchester City. In addition to that, the results of Manchester United are very similar to those of Manchester City against the teams on LEVEL 2 (4W, 1D, 1L of United compared to 5W, 0D, 1L of City).

Interestingly, even though Chelsea is fifth in the league table, it is placed in LEVEL 3. Although it might seem odd at a first glance, it makes sense at a closer look. The goal differences in direct comparison of Chelsea against the teams in the top three levels are:

- (LEVEL 1) Manchester City: −2, Manchester United: 0
- (LEVEL 2) Liverpool: +1, Arsenal: 0, Tottenham: −1
- (LEVEL 3) Leicester: +1, Newcastle: −1, West Ham: −1, Burnley: 0, Bournemouth: −2, Crystal Palace: 0, Watford: −1, Everton: +2

In the direct comparison, Chelsea did not win against a single team on LEVEL 1, only one team on LEVEL 2, and only two (out of eight) teams on LEVEL 3. This also reflects their inconsistency throughout last season. As for the teams in LEVEL 4, they are also in the bottom of the league table, and share similar overall statistics.

Serie A. As in the previous two leagues, the ILP discovers four levels in Serie A. Although Roma is above Inter in the league table, only Inter makes it to the top level. The goal differences in direct comparison of Inter against the teams in the top two levels are as follows:

Team	P	W	D	L	GD	Pts
Juventus	38	30	5	3	62	95
Napoli	38	28	7	3	48	91
Roma	38	23	8	7	33	77
Inter	38	20	12	6	36	72
Lazio	38	21	9	8	40	72
Milan	38	18	10	10	14	64
Atalanta	38	16	12	10	18	60
Fiorentina	38	16	9	13	8	57
Torino	38	13	15	10	8	54
Sampdoria	38	16	6	16	-4	54
Sassuolo	38	11	10	17	-30	43
Genoa	38	11	8	19	-10	41
Chievo	38	10	10	18	-23	40
Udinese	38	12	4	22	-15	40
Bologna	34	11	6	21	-12	39
Cagliari	34	11	6	21	-28	39
Spal	34	8	14	16	-20	38
Crotone	34	9	8	21	-26	35
Verona	34	7	4	27	-48	25
Benevento	34	6	3	29	-51	21

Rank	Teams
1	Juventus, Inter, Napoli
2	Roma, Torino, Lazio, Atalanta, Sampdoria, Milan, Fiorentina
3	Udinese, Crotone, Chievo, Spal, Bologna, Sassuolo, Genoa, Cagliari
4	Verona, Benevento

Team	P	W	D	L	GD	Pts
Barcelona	38	28	9	1	70	93
At Madrid	38	23	10	5	36	79
Real Madrid	38	22	10	6	50	76
Valencia	38	22	7	9	27	73
Villarreal	38	18	7	13	7	61
Betis	38	18	6	14	-1	60
Sevilla	38	17	7	14	-9	58
Getafe	38	15	10	13	9	55
Eibar	38	14	9	15	-6	51
Girona	38	14	9	15	-9	51
Espanyol	38	12	13	13	-6	49
Celta	38	13	10	15	-1	49
Sociedad	38	14	7	17	7	49
Alaves	38	15	2	21	-10	47
Levante	38	11	13	14	-14	46
Ath Bilbao	38	10	13	15	-8	43
Leganes	38	12	7	19	-17	43
La Coruna	38	6	11	21	-38	29
Las Palmas	38	5	7	26	-50	22
Malaga	38	5	5	28	-37	20

Rank	Teams
1	Barcelona
2	Real Madrid
3	Atlético Madrid, Sociedad, Villareal, Valencia
4	Celta, Eibar, Atlético Bilbao, Girona, Sevilla, Alaves, Leganes, Espanyol, Levante, Betis
5	Malaga, La Coruna, Las Palmas

Fig. 3. Results on (top) Serie A, and (bottom) La Liga. Accompanying the results on the right hand side are the corresponding league tables for the year 2017/2018.

- (LEVEL 1) Juventus: −1, Napoli: 0
- (LEVEL 2) Roma: +2, Lazio: +1, Milan: +1, Atalanta: +2, Fiorentina: +2, Torino −1, Sampdoria: +6

This clearly suggests that Inter had a very good record against the teams in the top two levels (except teams from Turin), and hence is placed in the top level.

La Liga. Contrary to the other leagues, the ILP discovers five levels in La Liga. Barcelona won La Liga last season with only one loss, and they were 14 points ahead of second-placed Atletico Madrid. It is sensible that they are the only team in LEVEL 1. Although Atletico Madrid is above Real Madrid in the league table (with mere 3 points difference), Real Madrid is the only team on LEVEL 2. If we look at last season's La Liga matches, we find that although Real Madrid has only marginally better performance against the top three teams in the league table compared to Atletico Madrid, Real Madrid did very well against the rest of the teams. The inclusion of Getafe on LEVEL 3 is due to its better head-to-head performances facing the teams on the same level.

6 Conclusion

We generalised hierarchies in weighted networks to sport leagues with draws. As a proof of concept, we considered four major European football leagues. The results show that the proposed algorithm yields a natural ranking of the teams in a football league into a small number of interpretable quality levels. Future work includes applying our generalised formulation of agony to other application areas, such as ranking the products of a retail store, and developing an efficient algorithms to discover hierarchies in those domains.

Acknowledgements. Stefan Neumann gratefully acknowledges the financial support from the Doctoral Programme "Vienna Graduate School on Computational Optimization" which is funded by the Austrian Science Fund (FWF, project no. W1260-N35). Kailash Budhathoki is supported by the International Max Planck Research School for Computer Science, Germany.

References

1. Football club elo ratings. http://clubelo.com. Accessed 03 Aug 2018
2. World football elo ratings. https://www.eloratings.net. Accessed 03 Aug 2018
3. Constantinou, A.C.: Dolores: a model that predicts football match outcomes from all over the world. Mach. Learn. **108**(1), 49–75 (2019). https://doi.org/10.1007/s10994-018-5703-7
4. Constantinou, A.C., Fenton, N.E.: Determining the level of ability of football teams by dynamic ratings based on the relative discrepancies in scores between adversaries. J. Quant. Anal. Sports **9**(1), 37–50 (2013)
5. Elo, A.E.: The Rating of Chess Players, Past and Present. Arco Pub., New York (1978)

6. Even, G., Naor, J.S., Schieber, B., Sudan, M.: Approximating minimum feedback sets and multi-cuts in directed graphs. In: Balas, E., Clausen, J. (eds.) IPCO 1995. LNCS, vol. 920, pp. 14–28. Springer, Heidelberg (1995). https://doi.org/10.1007/3-540-59408-6_38
7. Gupte, M., Shankar, P., Li, J., Muthukrishnan, S., Iftode, L.: Finding hierarchy in directed online social networks. In: WWW, pp. 557–566 (2011)
8. Hanauer, K.: Linear orderings of sparse graphs. Ph.D. thesis, University of Passau (2018)
9. Hanauer, K.: Linear orderings of sparse graphs supplementary (2018). https://sparselo.algo-rhythmics.org
10. Hubáček, O., Šourek, G., Železný, F.: Learning to predict soccer results from relational data with gradient boosted trees. Mach. Learn. 108(1), 29–47 (2018). https://doi.org/10.1007/s10994-018-5704-6
11. Hvattum, L.M., Arntzen, H.: Using ELO ratings for match result prediction in association football. Int. J. Forecast. 26(3), 460–470 (2010)
12. Karp, R.M.: Reducibility among combinatorial problems. In: Miller, R.E., Thatcher, J.W., Bohlinger, J.D. (eds.) Complexity of Computer Computations. IRSS, pp. 85–103. Springer, Boston (1972). https://doi.org/10.1007/978-1-4684-2001-2_9
13. Leitner, C., Zeileis, A., Hornik, K.: Forecasting sports tournaments by ratings of (prob)abilities: a comparison for the EURO 2008. Int. J Forecast. 26(3), 471–481 (2010)
14. Skinner, G.K., Freeman, G.H.: Soccer matches as experiments: how often does the 'best' team win? J. Appl. Stat. 36(10), 1087–1095 (2009). https://doi.org/10.1080/02664760802715922
15. Tatti, N.: Faster way to agony: discovering hierarchies in directed graphs. In: Calders, T., Esposito, F., Hüllermeier, E., Meo, R. (eds.) ECML PKDD 2014. LNCS (LNAI), vol. 8726, pp. 163–178. Springer, Heidelberg (2014). https://doi.org/10.1007/978-3-662-44845-8_11
16. Tatti, N.: Hierarchies in directed networks. In: ICDM, pp. 991–996 (2015)
17. Tatti, N.: Tiers for peers: a practical algorithm for discovering hierarchy in weighted networks. Data Min. Knowl. Discov. 31(3), 702–738 (2017)

US Team Sports

Interpreting Deep Sports Analytics: Valuing Actions and Players in the NHL

Guiliang Liu[✉], Wang Zhu, and Oliver Schulte

Simon Fraser University, Burnaby, Canada
{gla68,zhuwangz}@sfu.ca, oschulte@cs.sfu.ca

Abstract. Deep learning has started to have an impact on sports analytics. Several papers have applied action-value Q learning to quantify a team's chance of success, given the current match state. However, the black-box opacity of neural networks prohibits understanding why and when some actions are more valuable than others. This paper applies interpretable Mimic Learning to distill knowledge from the opaque neural net model to a transparent regression tree model. We apply Deep Reinforcement Learning to compute the Q function, and action impact under different game contexts, from 3M play-by-play events in the National Hockey League (NHL). The impact of an action is the change in Q-value due to the action. The play data along with the associated Q functions and impact are fitted by a mimic regression tree. We learn a general mimic regression tree for all players, and player-specific trees. The transparent tree structure facilitates understanding the general action values by feature influence and partial dependence plots, and player's exceptional characteristics by identifying player-specific relevant state regions.

1 Introduction

A fundamental goal of sports statistics is to quantify how much physical player actions contribute to winning in what situation. The advancement of sequential deep learning opens new opportunities of modeling complex sports dynamics, as more and larger play-by-play datasets for sports events become available. Action values based on deep learning are a very recent development, that provides a state-of-the-art player ranking method [12]. Several very recent works [10, 26] have built deep neural networks to model players' actions and value them under different situation. Compared with traditional statistics-based methods for action values [14, 19], deep models support a more comprehensive evaluation because (1) deep neural networks generalize well to different actions and complex game contexts and (2) various network structures (e.g. LSTM) can be applied to model the current game context and its sequential game history.

However, a neural network is an opaque black-box model. It prohibits understanding when or why the player's action is valuable, and which context features are the most influential for this assessment. A promising approach to overcome this limitation is *Mimic Learning* [1], which applies a transparent model to distill

© Springer Nature Switzerland AG 2019
U. Brefeld et al. (Eds.): MLSA 2018, LNAI 11330, pp. 69–81, 2019.
https://doi.org/10.1007/978-3-030-17274-9_6

the knowledge form the opaque model to an interpretable data structure. In this work, we train a Deep Reinforcement Learning (DRL) model to learn *action-value functions*, also known as Q functions, which represent a team's chance of success in a given match context. The *impact* of an action is computed as the difference between two consecutive Q values, before and after the action. To obtain an interpretable model that mimics the Q-value network, we first learn general regression trees for all players, for both Q and impact functions. The results show our trees achieve good mimic performance (small mean square error and variance). To understand the Q functions and impact, we compute the feature importance and use partial dependence plot to analyze the influence of different features with the mimic trees.

To highlight the strengths and weaknesses of an individual player compared to a general player, we construct player-specific mimic trees for Q values and impact. Based on a player-specific tree, we define an interpretable measure for which players are most exceptional overall.

Contribution. The main contributions of our paper are as follows: (1) A Mimic Learning Framework to interpret the action values from a deep neural network model. (2) Both a general mimic model and a player specific mimic model are trained and compared to find the influential features and exceptional players.

The paper is structured as follow: Sect. 2 covers the related work about the player evaluation metrics, Deep Sport Analytics and Interpretable Mimic Learning. Section 3 explains the reinforcement learning model of play dynamics from NHL dataset. Section 4 introduces the procedure of learning the Q values and Impact with DRL model, which completes our review of previous work. We show how to mimic DRL with regression tree in Sects. 5 and 6 discuss the interpretability of Q functions and Impact with Mimic tree. We highlight some exceptional players with the Mimic tree in Sect. 7.

2   Related Work

We discuss the previous work most related to our work.

Player Evaluation Metrics. Numerous metrics have been proposed to measure the players' performance.

One of the most common is *Plus-Minus (±)* [14], which measures how the presence of a player influences the goals of his team. But it considers only goals, and for context only which players are on the ice. *Total Hockey Rating (THoR)* [18] is an alternative metric that evaluates all the actions by whether or not a goal occurred in the following 20 s. Using a fixed time window rather makes this approach less useful for low-scoring goals like hockey and soccer. *Expected Possession Value (EPV)* [2] is an alternative metric, developed for basketball, that evaluate all players' actions by the points that they are expected to score. A POINTWISE Markov model is built to compute the point values with the spatial-temporal tracking data of players' state and actions. Many recent works have applied the *Reinforcement Learning (RL)* to compute a Q value to evaluate players actions. [12,17,19,20] built an Markov Decision Model from the sequential video tracking data and applied dynamic programming to learn the Q-functions. Value-above-replacement evaluates how many expected goals or wins the presence of a player adds, compared to a random player, giving rise to the GAR and WAR metrics [7]. Liu and Schulte [7] provide evidence that the Q-value ranking performs better than the GAR and WAR metrics.

Sport Analytics with Deep Models. Modelling sports dynamics with deep sequential neural nets is a rising trend [10,15]. Dynamical models predict the next event but do not evaluate the expected success from actions, as Q functions do. DRL for learning sports Q functions is a very recent topic [12,26]. Although these deep models provide an accurate evaluation of player actions, it is hard to understand why the model assigns a large influence to a player in a given situation.

Interpretable Mimic Learning. Complex deep neural networks are hard to interpret. An approach to overcome this limitation is Mimic Learning [1]. Recent works [3,4] have demonstrated that simple models like shallow feed-forward neural network or decision trees can mimic the function of a deep neural network. Soft outputs are collected by passing inputs to a large, complex and accurate deep neural network. Then we train a mimic model with the same input and soft output as supervisor. The results indicate that training a mimic model with soft output achieves substantial improvement in accuracy and efficiency, over training the same model type directly with hard targets from the dataset.

Table 1. Dataset example   **Table 2.** Derived features

GID	PID	GT	TID	X	Y	MP	GD	Action	OC	P
1365	126	14.3	6	−11.0	25.5	Even	0	Lpr	S	A
1365	126	17.5	6	−23.5	−36.5	Even	0	Carry	S	A
1365	270	17.8	23	14.5	35.5	Even	0	Block	S	A
1365	126	17.8	6	−18.5	−37.0	Even	0	Pass	F	A
1365	609	19.3	23	−28.0	25.5	Even	0	Lpr	S	H
1365	609	19.3	23	−28.0	25.5	Even	0	Pass	S	H

Velocity	TR	D	Angle	H/A	PN
(−23.4, 1.5)	3585.7	3.4	0.250	A	4
(−4.0, −3.5)	3582.5	3.1	0.314	A	4
(−27.0, −3.0)	3582.2	0.3	0.445	H	4
(0, 0)	3582.2	0.0	0.331	A	4
(−30.3, −7.5)	3580.6	1.5	0.214	H	5
(0, 0)	3580.6	0.0	0.214	H	5

GID = GameId, PID = playerId, GT = GameTime, TID = TeamId, MP = Manpower, GD = Goal Difference, OC = Outcome, S = Succeed, F = Fail, P = Team Possess puck, H = Home, A = Away, TR = Time Remain, PN = Play Number, D = Duration.

3   Play Dynamics in NHL

Dataset. The Q-function approach was originally developed using the publicly available NHL data [17]. Our deep RL model could be applied to this data, but in this paper, we utilize a richer proprietary dataset constructed by SPORT-LOGiQ with computer vision techniques. It provides information about *game events* and *player actions* for the entire 2015–2016 NHL season, which contains

3,382,129 events, covering 30 teams, 1,140 games and 2,233 players. Table 1 shows an excerpt. The data tracks events around the puck, and record the identity and actions of the player, with space and time stamps, and features of the game context. The unit for space stamps are feet and for time stamps seconds. We utilize adjusted spatial coordinates, where negative numbers refer to the defensive zone of the acting player, positive numbers to his offensive zone. Adjusted X-coordinates (XAdjcoord) run from -100 to $+100$, Y-coordinates (YAdjcoord) from 42.5 to -42.5, and the origin is at the ice center. We include data points from all manpower scenarios, not only even-strength, and add the manpower context as a feature. We did not include overtime data. Period information is implicitly represented by game time. We augment the data with derived features in Table 2 and list the complete feature set in Table 3.

Table 3. Complete feature list. Values for the feature Manpower are EV = Even Strength, SH = Short Handed, PP = Power Play.

Name	Type	Range
X Coordinate of puck	Continuous	$[-100, 100]$
Y Coordinate of puck	Continuous	$[-42.5, 42.5]$
Velocity of puck	Continuous	$(-\inf, +\inf)$
Time remaining	Continuous	$[0, 3600]$
Score differential	Discrete	$(-\inf, +\inf)$
Manpower	Discrete	{EV, SH, PP}
Event duration	Continuous	$[0, +\inf)$
Action outcome	Discrete	{successful, failure}
Angle between puck and goal	Continuous	$[-3.14, 3.14]$
Home/Away team	Discrete	{Home, Away}

Reinforcement Learning Model. Our notation for RL concepts follows [17]. There are two agents *Home* resp. *Away* representing the home and away team, respectively. The **reward**, represented by goal vector $\mathbf{g_t}$, is a 1-of-3 indicator vector that specifies which team scores. For readability, we use *Home, Away, Neither* to denote the team in a goal vector (e.g. $g_{t,Home} = 1$ means that the home team scores at time t). An **action** a_t is one of 13 types, including shot, assist, etc., with a mark that specifies the team executing the action, e.g. *Shot(Home)*. An observation is a feature vector $\mathbf{x_t}$ for discrete time step t specifies a value for the 10 features listed in Table 3. A **sequence** s_t is a list $(x_0, a_0, \ldots, x_t, a_t)$ of observation-action pairs.

We divide NHL games into **goal-scoring episodes**, so that each episode (1) begins at the beginning of the game, or immediately after a goal, and (2) terminates with a goal or the end of the game. We define a Q function to represent the conditional probability of the event that the home resp. away team *scores* *the goal at the end of the current goal-scoring episode* (denoted $goal_{Home} = 1$ resp. $goal_{Away} = 1$), or neither team does (denoted $goal_{Neither} = 1$):

$$Q_{team}(s, a) = P(goal_{team} = 1|s_t = s, a_t = a).$$

4   Q-Values and Action Impact

We review learning Q values and impact, using neural network Q-function approximation. A Tensorflow script is available on-line [13].

4.1   Compute Q Functions with Deep Reinforcement Learning

We apply the on policy Temporal Difference (TD) prediction method Sarsa [22] to estimate $Q_{team}(s, a)$ for current policies π_{home} and π_{away}. The neural network has three fully connected layers connected by a ReLu activation function. The number of input nodes equals the sum of the dimensions of feature vector **s** and action vector **a**. The number of output nodes is three, including $\hat{Q}_{Home}$, $\hat{Q}_{Away}$ and $\hat{Q}_{Neither}$, which are normalized to probability. The parameters θ of neural network are updated by minibatch gradient descent with optimization method Adam. Using mean squared error function, the Sarsa Gradient Descent at training step i is based on the square of TD error:

$$\mathcal{L}(\theta_i) = 1/B \sum_t^B (g_t + \hat{Q}(s_{t+1}, a_{t+1}, \theta_i) - \hat{Q}(s_t, a_t, \theta_i))^2$$

$$\theta_t = \theta_t + \alpha \nabla_\theta \mathcal{L}(\theta_t)$$

where B is the batch size and α is the learning rate optimized by the Adam algorithm [8]. For post-hoc interpretability [11] for the learned Q function, we illustrate its temporal and spatial projections in Figs. 1 and 2.

Temporal Projection. Figure 1 plots a value ticker [6] that represents the evolution of the action-value Q function (including Q values for home, away team and neither) from the 3rd period of a randomly selected match between Penguins (Home) and Canadians (Away), Oct.13, 2015. Sports analysts and commentators use ticker plots to highlight critical match events [6]. We mark significant changes in the scoring probabilities and their corresponding events.

Spatial Projection. The neural network generalizes from observed sequences and actions to sequences and actions that have not occurred in our dataset. So we plot the learned smooth value surface $\hat{Q}^{Home}(s_\ell, shot(team))$ over the entire rink for home team shots in Fig. 2. Here s_ℓ represents the average play history for a shot at location ℓ, which runs in unit steps from $x\_axis \in [-100, 100]$ and $y\_axis \in [-42.5, 42.5]$. It can be observed that (1) The chance that the home team scores after a shot is shown to depend on the angle and distance to the goal. (2) Action-value function generalizes to the regions where shots are rarely observed (At the lower or upper corner of the rink).

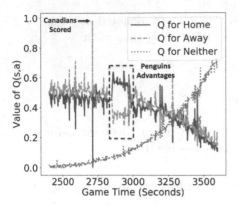
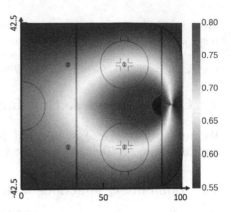

Fig. 1. Temporal projection: evolution of scoring probabilities for the next goal, including the chance that neither team scores another goal.

Fig. 2. Spatial projection for the shot action: the probability that the home team scores the next goal after taking a shot at a rink location, averaged over possible game states.

4.2 Evaluate Players with Impact Metric

We follow previous work [12] and evaluate players by how much their actions change the expected return of their team's in a given game state [12]. This quantity is defined as the **Impact** of an action under current environment (observation) s_t. Players' overall performance can be estimated by summing the impact of players throughout a game season. The resulting metric is named **Goal Impact Metric (GIM)**.

$$impact^{team}(s_t, a_t) = \hat{Q}^{team}(s_t, a_t) - \hat{Q}^{team}(s_{t-1}, a_{t-1})$$
$$GIM^i(D) = \sum_{s,a} n_D^i(s, a) \times impact^{team_i}(s, a)$$

Table 4 shows the top 10 players ranked by GIM. Our purpose in this paper is to interpret the Q values and the impact ranking, not to evaluate them. Previous work provides extensive evaluation [12,17,19,20]. We summarize some of the main points. (1) The impact metric passes the "eye test". For example the players in Table 4 are well-known top performers. (2) The metric correlates strongly with various quantities of interest in the NHL, including goals, points, Time-on-Ice, and salary. (3) The metric is consistent between and within seasons. (4) The impact is assessed for *all* actions, including defensive and offensive actions. It therefore not biased towards forwards. For instance, defenceman Erik Karlsson appears at the top of the ranking.

5 Mimicking DRL with Regression Tree

We apply Mimic Learning [1] and train a transparent regression tree to mimic the black-box neural network. As it is shown in Fig. 3, our framework aims at

Table 4. 2015–2016 Top-10 player impact scores

Name	GIM	Assists	Goals	Points	±	Age	Team	Salary
Taylor Hall	96.40	39	26	65	−4	24	EDM	$6,000,000
Joe Pavelski	94.56	40	38	78	25	31	SJS	$6,000,000
Johnny Gaudreau	94.51	48	30	78	4	22	CGK	$925,000
Anze Kopitar	94.10	49	25	74	34	28	LAK	$7,700,000
Erik Karlsson	92.41	66	16	82	−2	25	OTT	$7,000,000
Patrice Bergeron	92.06	36	32	68	12	30	BOS	$8,750,000
Mark Scheifele	90.67	32	29	61	16	23	WPG	$832,500
Sidney Crosby	90.21	49	36	85	19	28	PIT	$12,000,000
Claude Giroux	89.64	45	22	67	−8	28	PHI	$9,000,000
Dustin Byfuglien	89.46	34	19	53	4	31	WPG	$6,000,000

mimicking Q functions and impact. We first train the *general tree model* with the deep model's input/output for all players and then use it to initialize the *player-specific model* for an individual player (Sect. 7). The transparent tree structure provides much information for understanding the Q functions and impact.

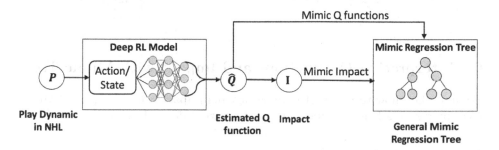

Fig. 3. Interpretable Mimic Learning Framework

We focus on two mimicking targets: *Q functions* and *Impact*. For *Q* functions, we fit the mimic tree with the NHL play data and their associated soft outputs (Q values) from our DRL model (neural network). The last 10 observations (determined experimentally) from the sequence are extracted, and CART regression tree learning is applied to fit the soft outputs. This is a multi-output regression task, as our DRL model outputs a Q vector containing three Q values ($\hat{Q}_t = \langle \hat{Q}_t^{home}, \hat{Q}_t^{away}, \hat{Q}_t^{end} \rangle$) for an observation features vectors (s_t) and an action (a_t). A straightforward approach for the multi-target regression problem is training a separate regression model for each Q value. But separate trees for each Q function are somewhat difficult to interpret. An alternative approach to reduce the total tree size is training a Multi-variate Regression Tree (MRTs) [5], which fits all three Q values simultaneously in a regression tree. An MRT can also model the dependencies between the different Q variables [21]. For Impacts, we have only one output ($impact_t$) for each sequence (s_t) and current action (a_t) at time step t.

We examine the mimic performance of regression tree for the Q functions and impact. A common problem of regression trees is over-fitting. We use the Mean Sample Leaf (MSL) to control the minimum number of samples at each leaf node. We apply ten-fold cross validation to measure the performance of our mimic regression tree by Mean Square Error (MSE) and variance. As is shown in Table 5, the tree achieves satisfactory performance when MSL equals 20 (the minimum MSE for Q functions, small MSE and variance for impact).

Table 5. Performance of General Mimic Regression Tree (RT) with different Minimum Samples in each Leaf node (MSL). We apply ten-fold cross validation and report the regression result with format: Mean Square Error (Variance)

Model	Q_home	Q_away	Q_end	Impact
RT-MSL1	3.35E-04(1.43E-09)	3.21E-04(1.26E-09)	1.74E-04(2.18E-09)	1.33E-03(5.43E-09)
RT-MSL5	2.59E-04(1.07E-09)	2.51E-04(0.89E-09)	1.35E-04(1.87E-10)	9.84E-04(2.72E-09)
RT-MSL10	2.38E-04(1.02E-09)	2.30E-04(0.89E-09)	1.25E-04(2.30E-10)	8.66E-04(2.17E-09)
RT-MSL20	**2.31E-04(0.92E-09)**	**2.22E-04(0.82E-09)**	**1.23E-04(2.05E-10)**	7.92E-04(1.45E-09)
RT-MSL30	2.35E-04(0.98E-09)	2.27E-04(0.85E-09)	1.27E-04(2.32E-10)	7.67E-04(1.16E-09)
RT-MSL40	2.39E-04(0.96E-09)	2.30E-04(0.85E-09)	1.29E-04(2.19E-10)	**7.58E-04(1.10E-09)**

6   Interpreting Q Functions and Impact with Mimic Tree

We now show how to interpret Q functions and Impact using the general Mimic tree, by deriving feature importance and a partial dependence plot.

6.1   Compute Feature Importance

In CART regression tree learning, variance reduction is the criterion for evaluating the quality of a split. Therefore we compute the importance of a target feature by summing the variance reductions at each split using the target feature [3]. We list the top 10 important features in the mimic tree for Q values and impact in Table 6. The frequency of a feature is the number of times the tree splits on the feature. The notation $T - n : f$ indicates that a feature occurs n time steps before the current time. We find that the Q and impact functions agree on nearly half of the features, but their importance values differ. For Q values, time remaining is the most influence features with significantly larger importance value than other. This is because less time means fewer chance of any goals (see Fig. 1). But for impact, time remaining is much less important, because impact is the difference of consecutive Q values, which cancels the time effect and focuses only on the influence of a player's action a: Near the end of the match, players still have a chance to make actions with high impact. The top three important features for impact are (1) Goal: if the player scores a goal. (2) Shot-on-Goal Outcome: if the player's shot is on target (3) X Coordinate: the x-location of the puck (goal-to-goal axis). Thus the impact function recognizes players for shooting, successful actions, and for advancing the puck towards the goal of their opponent. A less intuitive finding is that the duration of an action affects its impact. Notice that for both Q values and impact, the top ten important features contain historical features (with $T - n$ for $n > 0$), which supports the importance of including historical data in observation sequence s.

Table 6. Top 10 features for Q values (left) and Impact (right). The notation $T - n : f$ indicates that a feature occurs n time steps before the current time.

Feature Name	Frequency	Importance	Feature Name	Frequency	Importance
T: Time Remaining	12,524	0.817431	T: Goal	1	0.160595
T-1: Manpower	93	0.070196	T: Shot-on-Goal	1	0.099482
T-1: Team Identifier	57	0.020504	T: X Coordinate	7,142	0.077410
T: Manpower	346	0.017306	T: X Velocity	8,087	0.041903
T: Shot	31	0.011159	T-1: X Coordinate	3,591	0.041847
T: Score Differential	3,229	0.009568	T: Angle to Goal	7,525	0.041607
T: X Coordinate	11,797	0.006968	T: Time Remaining	8,669	0.036289
T-1: X Coordinate	3,406	0.006963	T: Duration	7,411	0.028831
T-2: Manpower	82	0.005045	T: Home/Away Team	378	0.027177
T: Home/Away Team	135	0.003755	T: Y Coordinate	6,890	0.027597

6.2 Draw Partial Dependence Plot

A partial dependence plot is a common visualization to determine qualitatively what a model has learned and thus provides interpretability [3,11]. The plot approximates the prediction function for a single target feature, by marginalizing over the values of all other features. We select X Coordinate (of puck), Time Remaining and X Velocity (of puck), three continuous features with high importance for both the Q and the impact mimic tree. As it is shown in Fig. 4, Time Remaining has significant influence on Q values but very limited effect on impact. This is consistent with our findings for feature importance. For X Coordinate, as a team is likely to score the next goal in the offensive zone, both Q values and impact increase significantly when the puck is approaching its opponent's goal (larger X Coordinate). And compared to the position of the puck, velocity in X-axis has limited influence on Q values but it does affect the impact. This shows that the impact function uses speed on the ice as an important criterion for valuing a player. We also observe the phenomenon of home advantage [23] as the Q value (scoring probability) of the home team is slightly higher than that of the away team.

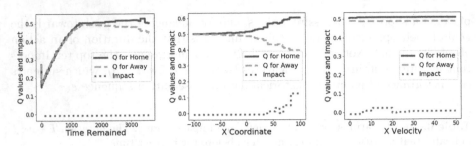

Fig. 4. Partial dependence plot for Time Remaining (left), X Coordinate (middle) and X Velocity (right)

7   Highlighting Exceptional Players

Our approach to quantifying which players are exceptional is based on a partition the continuous state space into a discrete set of m disjoint regions. Given a Q or impact function, exceptional players can be found by region-wise comparison of a player's excepted impact to that of a random player's. For a specific player, this comparison highlights match settings in which the player is especially strong or weak. The formal details are as follows.

Let n_D be the number of actions by player P, of which n_ℓ fall into discrete state region $\ell = 1, \ldots, m$. For a function f, let $\hat{f}_\ell$ be the value of f estimated from *all* data points that fall into region ℓ, and let $\hat{f}_\ell^P$ be the value of f estimated from the n_ℓ data points for region ℓ and player P. Then the weighted squared f-difference is given by:

$$\sum_\ell n_\ell / n_D (\hat{f}_\ell - \hat{f}_\ell^P)^2. \tag{1}$$

Regression trees provide an effective way to discretize a Q-function for a continuous state space [25]: Each leaf forms a partition cell in state space (constructed by the splits with various features along the path from root to the leaf). The regression trees described in Sect. 5 could be used, but they represent *general* discretizations learned for all the players over a game season, which means that they may miss distinctions that are important for a specific player. For example, if an individual player is especially effective in the neutral zone, but the average player's performance is not special in the neutral zone, the generic tree will not split on "neutral zone" and therefore will not be able to capture the individual's special performance. Therefore we learn for each player, a *player-specific* regression tree.

The General Tree is learned with all the inputs and their corresponding Q or Impact values (soft labels). The Player Tree is initialized with the General Tree and then fitted with the n_D datapoints of a specific player P and their corresponding Q values ($f_{\hat{Q}}^P(f_{\hat{Q}}, s_t^P, a_t^P) \rightarrow range(\hat{Q}_t^P)$) or Impact values ($f_I^P(f_I, s_t^P, a_t^P) \rightarrow range(Impact_t^P)$). It inherits the tree structure of the general model RT-MSL20 in Sect. 5, uses the target player data to prune the general tree, then expands the tree with further splits. Initializing with the general tree assumes players share relevant features and prevents over-fitting to a player's specific data. A Player Tree defines a discrete set of state regions, so we can apply Eq. 1 with the Q or impact functions. Table 7 shows the weighted squared differences for the top 5 players in the GIM metric.

Table 7. Exceptional players based on tree discretization

Player	Q_home	Q_away	Q_end	Impact
Taylor Hall	1.80E-04	2.49E-04	2.28E-04	6.66E-05
Joe Pavelski	**4.64E-04**	**2.90E-04**	**3.04E-04**	1.09E-04
Johnny Gaudreau	2.12E-04	1.96E-04	1.43E-04	6.77E-05
Anze Kopitar	2.58E-04	2.00E-04	2.43E-04	8.28E-05
Erik Karlsson	2.97E-04	1.89E-04	1.86E-04	**2.00E-04**

We find that (1) Joe Pavelski, who scored the most in the 2015–2016 game season, has the largest Q values difference and (2) Erik Karlsson, who had the most points (goal+assists), has the largest Impact difference. They are the two players who differ the most from the average players by Q-value and Impact.

8   Conclusion and Future Work

This paper applies Mimic Learning to understand the Q function and impact from Deep Reinforcement Learning Model in valuing actions and players. To study the influence of a feature, we analyze a general mimic model for all players by feature importance and partially dependence plot. For individual players, performance in state regions defined by the player specific tree is implemented to find exceptional players. With our interpretable Mimic Learning, coaches and fans can understand what the deep models have learned and thus trust the results. While our evaluation focuses on ice hockey, our techniques apply to other continuous-flow sports such as soccer and basketball.

In *future work*, the player trees can be used to highlight match scenarios where a player shows exceptionally strong or weak performance, in both defense and offense. A limitation of our current model is that it pools all data from the different teams, rather than modelling the differences among teams. A hierarchical model for ice hockey can be used to analyze how teams are similar and how they are different, like those that have been built for other sports (e.g., cricket [16].) Another limitation is that players get credit only for recorded individual actions. An influential approach to extend credit to all players on the rink has been based on regression [9,14,24]. A promising direction for future work is to combine Q-values with regression.

References

1. Ba, J., Caruana, R.: Do deep nets really need to be deep? In: Advances in Neural Information Processing Systems, pp. 2654–2662 (2014)
2. Cervone, D., D'Amour, A., Bornn, L., Goldsberry, K.: Pointwise: predicting points and valuing decisions in real time with NBA optical tracking data. In: Proceedings of the 8th MIT Sloan Sports Analytics Conference, Boston, MA, USA, vol. 28, p. 3 (2014)
3. Che, Z., et al.: Interpretable deep models for ICU outcome prediction. In: AMIA Annual Symposium Proceedings, vol. 2016, p. 371. AMIA (2016)
4. Dancey, D., Bandar, Z.A., McLean, D.: Logistic model tree extraction from artificial neural networks. IEEE Trans. Syst. Man Cybern. Part B (Cybern.) 37(4), 794–802 (2007)
5. De'Ath, G.: Multivariate regression trees: a new technique for modeling species-environment relationships. Ecology 83(4), 1105–1117 (2002)
6. Decroos, T., Dzyuba, V., Van Haaren, J., Davis, J.: Predicting soccer highlights from spatio-temporal match event streams. In: AAAI, pp. 1302–1308 (2017)
7. Gerstenberg, T., Ullman, T., Kleiman-Weiner, M., Lagnado, D., Tenenbaum, J.: Wins above replacement: Responsibility attributions as counterfactual replacements. In: Proceedings of the Annual Meeting of the Cognitive Science Society, vol. 36 (2014)
8. Glorot, X., Bordes, A., Bengio, Y.: Deep sparse rectifier neural networks. In: Proceedings of the Fourteenth International Conference on Artificial Intelligence and Statistics, pp. 315–323 (2011)
9. Kharrat, T., Pena, J.L., McHale, I.: Plus-minus player ratings for soccer. arXiv preprint arXiv:1706.04943 (2017)
10. Le, H.M., Carr, P., Yue, Y., Lucey, P.: Data-driven ghosting using deep imitation learning. In: MIT Sloan Sports Analytics Conference (2017)
11. Lipton, Z.C.: The mythos of model interpretability. arXiv preprint arXiv:1606.03490 (2016)
12. Liu, G., Schulte, O.: Deep reinforcement learning in ice hockey for context-aware player evaluation. In: Proceedings IJCAI-18, pp. 3442–3448, July 2018. https://doi.org/10.24963/ijcai.2018/478
13. Liu, G., Schulte, O.: Drl-ice-hockey (2018). https://github.com/Guiliang/DRL-ice-hockey
14. Macdonald, B.: A regression-based adjusted plus-minus statistic for NHL players. J. Quant. Anal. Sports 7(3), 29 (2011)
15. Mehrasa, N., Zhong, Y., Tung, F., Bornn, L., Mori, G.: Deep learning of player trajectory representations for team activity analysis. In: MIT Sloan Sports Analytics Conference (2018)
16. Perera, H., Davis, J., Swartz, T.: Assessing the impact of fielding in twenty20 cricket. J. Oper. Res. Soc. 69, 1335–1343 (2018)
17. Routley, K., Schulte, O.: A Markov game model for valuing player actions in ice hockey. In: Proceedings Uncertainty in Artificial Intelligence (UAI), pp. 782–791 (2015)
18. Schuckers, M., Curro, J.: Total hockey rating (THoR): a comprehensive statistical rating of national hockey league forwards and defensemen based upon all on-ice events. In: 7th Annual MIT Sloan Sports Analytics Conference (2013)
19. Schulte, O., Khademi, M., Gholami, S., Zhao, Z., Javan, M., Desaulniers, P.: A Markov game model for valuing actions, locations, and team performance in ice hockey. Data Min. Knowl. Discovery 31(6), 1735–1757 (2017)

20. Schulte, O., Zhao, Z., Javan, M., Desaulniers, P.: Apples-to-apples: clustering and ranking NHL players using location information and scoring impact. In: MIT Sloan Sports Analytics Conference (2017)
21. Struyf, J., Džeroski, S.: Constraint based induction of multi-objective regression trees. In: Bonchi, F., Boulicaut, J.-F. (eds.) KDID 2005. LNCS, vol. 3933, pp. 222–233. Springer, Heidelberg (2006). https://doi.org/10.1007/11733492_13
22. Sutton, R.S., Barto, A.G.: Introduction to Reinforcement Learning, vol. 135. MIT Press, Cambridge (1998)
23. Swartz, T.B., Arce, A.: New insights involving the home team advantage. Int. J. Sports Sci. Coaching 9(4), 681–692 (2014)
24. Thomas, A., Ventura, S., Jensen, S., Ma, S.: Competing process hazard function models for player ratings in ice hockey. Ann. Appl. Stat. 7(3), 1497–1524 (2013). https://doi.org/10.1214/13-AOAS646
25. Uther, W.T., Veloso, M.M.: Tree based discretization for continuous state space reinforcement learning. In: AAAI/IAAI, pp. 769–774 (1998)
26. Wang, J., Fox, I., Skaza, J., Linck, N., Singh, S., Wiens, J.: The advantage of doubling: a deep reinforcement learning approach to studying the double team in the NBA. In: MIT Sloan Sports Analytics Conference (2018)

Player Pairs Valuation in Ice Hockey

Dennis Ljung, Niklas Carlsson, and Patrick Lambrix[✉]

Linköping University, Linköping, Sweden
patrick.lambrix@liu.se

Abstract. To overcome the shortcomings of simple metrics for evaluating player performance, recent works have introduced more advanced metrics that take into account the context of the players' actions and perform look-ahead. However, as ice hockey is a team sport, knowing about individual ratings is not enough and coaches want to identify players that play particularly well together. In this paper we therefore extend earlier work for evaluating the performance of players to the related problem of evaluating the performance of player pairs. We experiment with data from seven NHL seasons, discuss the top pairs, and present analyses and insights based on both the absolute and relative ice time together.

Keywords: Sports analytics · Data mining · Player valuation

1 Introduction

In the field of sports analytics, many works focus on evaluating the performance of players. A commonly used method to do this is to attribute values to the different actions that players perform and sum up these values every time a player performs these actions. These summary statistics can be computed over, for instance, games or seasons. In ice hockey, common summary metrics include the number of goals, assists, points (assists + goals) and the plus-minus statistics $(+/-)$, in which 1 is added when the player is on the ice when the player's team scores (during even strength play) and 1 is subtracted when the opposing team scores (during even strength). More advanced measures are, for instance, Corsi and Fenwick[1].

However, these metrics do not capture the context of player actions and the impact they have on the outcome of later actions. To address this shortcoming and to capture the ripple effect of actions (where one action increases/decreases the success of a later action, for example), recent works [6,11,13] have therefore introduced more advanced metrics that take into account the context of the actions and perform look-ahead. The use of look-ahead is particularly valuable in low-scoring sports such as ice hockey.

An important aspect that the above works do not take into account is that ice hockey is a team sport. In particular, individual ratings are not enough to predict

---

[1] See, e.g., https://en.wikipedia.org/wiki/Analytics_(ice_hockey).

© Springer Nature Switzerland AG 2019
U. Brefeld et al. (Eds.): MLSA 2018, LNAI 11330, pp. 82–92, 2019.
https://doi.org/10.1007/978-3-030-17274-9_7

outcomes for the team. It is therefore important for coaches to identify players that play particularly well together. In this paper, we therefore extend the above recent analysis approach to the related problem of evaluating the performance of **pairs of players**. For our analysis, we extend the work by Routley and Schulte [11] to evaluate the performance of player pairs. More specifically, we use their action-value Q-function to assign values to individual actions performed by players and then sum the value-adding actions associated with player pairs when they are on the ice simultaneously.

In ice hockey there are usually two defenders, three forwards, and a goaltender on the ice simultaneously. However, the set of players changes frequently (e.g., the average shift duration is roughly 45 s) and coaches typically select to adjust which players are on the ice at each point in time based on desirable matchups against the other teams and, perhaps most importantly, based on which players play well together. While ice hockey coaches traditionally talk about defense pairings (consisting of two defenders) and forward lines (consisting of three forwards), coaches increasingly work with both defense pairings and forward pairings. Identifying forwards pairs with particularly good "chemistry" is considered simpler and allows some flexibility when there are injuries on a team, for example. By focusing on pairings, we provide a tool that helps identify player pairs that perform particularly good/bad.

For our analysis, we use this tool to identify particularly successful pairings, compare successful pairings to how their coaches assign ice time to these pairings, and to compare the success of different categories of player pairs (e.g., based on position, total ice time together, or the relative fraction of their ice time played together). At a high level we find that coaches' desire to play their top players against the other teams' top players appears to even out the relative impact per minute observed across different player categories.

The remainder of the paper is organized as follows. Section 2 discusses related work, and Sect. 3 describes the work of [11]. In Sect. 4 we define our metrics to evaluate pairs of players and in Sect. 5 we present and discuss the results of our experiments with NHL play-by-play data from the 2007–2008 through 2013–2014 NHL seasons, as provided by [11]. The paper concludes in Sect. 6.

2   Related Work

Regarding player performance in ice hockey, several regression models have been proposed for dealing with the weaknesses of the +/− measure (e.g., [4,7,8]). Other measures have also been introduced, including Corsi, Fenwick, and added goal value [10].

Another measure for player evaluation based on the events that happen when a player is on the ice is proposed in [12]. Event impacts are based on the probability that the event leads to a goal (for or against) in the next 20 s.

Other works model the dynamics of an ice hockey game using Markov games where two opposing sides (e.g., the home team and the away team) try to reach states in which they are rewarded (e.g., scoring a goal). In [15] the scoring rate for each team is modeled as a semi-Markov process, with hazard functions for each process that depend on the players on the ice. A Markov win probability model given the goal and manpower differential state at any point in a hockey game is proposed in [5]. In [6,11,13,14] action-value Q-functions are learned with respect to different targets. (See Sect. 3 for the model in [11].) Although the approaches use Markov-based approaches, the definitions of states and reward functions are different. The advantages of such approaches (e.g., [14]) are the ability to capture game context (goals have different values in a tie game than in a game where a team is leading with many goals), the ability to look ahead and thereby assigning values to all actions in the game, and the possibility to define a player's impact through the player's actions. In this paper we base our work on one of these approaches, i.e., the work in [11].

There is not much work on evaluating player pair performance for ice hockey. In [15] player pairs are rated using their Markov-based model of the game. They selected the 1,000 player pairs with respect to the number of full-strength shifts in five NHL seasons (2007–2008 until 2011–2012) where the players in the pairs are both forwards or both defenders. Lessons learned consisted of knowledge about player pairs where the players performed better than their individual performance, and player pairs where the combination reduced the performance. Further, there is some work in basketball, such as [1,2] where player pairs are ranked according to a $+/-$ measure. Which types of players should be chosen for the top pair of players, is investigated in [3].

3 Background

In this section we explain the model of [11]. In [11] action-value Q-functions are learned with respect to the next goal or the next penalty. In this paper we only use the next goal Q-function.

The state space considers *action events* with three parameters: the action type (Faceoff, Shot, Missed Shot, Blocked Shot, Takeaway, Giveaway, Hit, Goal), the team that performs the action, and the zone (offensive, neutral, defensive).

A *play sequence* is defined as the empty sequence or a sequence of events for which the first event is a start marker, the possible next events are action events, and the possible last event is an end event. If the play sequence ends with an end event, it is a *complete* play sequence. The start/end events are Period Start, Period End, Early Intermission Start, Penalty, Stoppage, Shootout Completed, Game End, Game Off, and Early Intermission End.

Actions and play sequences occur in a context. In [11] a *context state* contains values for 3 context features. Goal Differential is the number of home goals minus the number of away goals. Manpower Differential is the number of home players on the ice minus the number of away players on the ice. Further, the Period of the game is recorded.

A *state* is then a pair which contains a context state and a (not necessarily complete) play sequence.

Actions are performed in specific states. For action a and state $s = <c, ps>$, where c is the context state and ps is the play sequence, the resulting state of performing a in state s is denoted by $s * a$ and is defined as $<c, ps * a>$, where $ps * a$ is the play sequence obtained by appending action a to ps. For states with play sequences that are end events, the next state is a state of the form $<c', \emptyset>$ where c' is defined by the end event. For instance, a goal will change the goal differential and update the context.

Transition probabilities between different states are based on play-by-play data. The transition probability $TP(s,s')$ for a transition from state s to state s' is defined as $Occ(s,s')/Occ(s)$ where $Occ(s)$ is the number of occurrences of s in the play-by-play data and $Occ(s,s')$ is the number of occurrences of s that are immediately followed by s' in the play-by-play data.

Using a state transition graph with the computed transition probabilities, Q-values for states are learned using a value iteration algorithm. The impact of an action in a certain state impact(s,a) is then defined as $Q_T(s * a) - Q_T(s)$ where T is the team performing the action.

The performance of a player is computed as the sum of the impacts of the actions the player performs (over a game or a season). This is equivalent to comparing the actions taken by a specific player to the actions of an average player.

For our work, we reimplemented the code available from [11] using Python and C++. The reward for goals for is $+1$ and goals against -1. The resulting impact values for the actions were used as a base for our work on pair performance.

4   Player Pair Metrics

We base our method for computing performance measures on the impact of actions[2] as defined by [11]. However, we define the impact of players in different ways. First, we define different sets of actions for players and player pairs (Table 1). We differentiate between actions performed by a player and actions performed (by the player or another player) when a player is on the ice. The player impact is then defined using the actions when the player is on the ice (Table 2). This allows for a measure that includes indirect impact on the game by being on the ice. Even when players do not perform registered actions, they can still influence the game; e.g., by opening up a path for a teammate who may score. Further, we define the direct impact of a player based on the actions the player performs (and this is essentially the impact as defined in [11]).

For player pairs we define the impact using the actions when both players are on the ice. To be able to measure the influence of the players in the pair on each other, we also define the impact of a player without a particular second player, i.e., we use the actions when the first player is on the ice, but not the second.

5   Data-Driven Analysis

For our analysis, we used NHL play-by-play data from the 2007–2008 through 2013–2014 NHL season, as provided by [11]. To compare against prior works,

---

[2] In the remainder we use action as a shorthand for action in a particular state.

Table 1. Basic action sets.

A is the set of all state-action-pairs $< s, a >$ where action a is performed in state s
$A_i(p_k)$ is the set of state-action-pairs when player p_k is on the ice
$A_i(p_k,p_l)$ is the set of state-action-pairs when players p_k and p_l both are on the ice
$\quad A_i(p_k,p_l) = A_i(p_k) \cap A_i(p_l)$
$A_p(p_k)$ is the set of state-action-pairs where the action is performed by player p_k
$\quad A_p(p_k) \subseteq A_i(p_k)$

Table 2. Player and player pair impact.

The direct impact of a player is the sum of the impact values of the actions performed by the player:
$\quad$ D-impact$(p_k) = \Sigma_{<s,a> \in A_p(p_k)}$ impact(s,a)
The impact of a player is the sum of the impact values of the actions when the player is on the ice:
$\quad$ impact$(p_k) = \Sigma_{<s,a> \in A_i(p_k)}$ impact(s,a)
The impact of a player pair is the sum of the impact values of the actions when both players are on the ice:
$\quad$ impact$(p_k,p_l) = \Sigma_{<s,a> \in A_i(p_k,p_l)}$ impact(s,a)
The impact of a player without a second player is the sum of the impact values of the actions when the player is on the ice and the second player is not on the ice:
$\quad$ impact-without$(p_k,p_l) = \Sigma_{<s,a> \in (A_i(p_k) \backslash A_i(p_k,p_l))}$ impact(s,a)

we primarily focus our analysis on the last two full regular seasons in this set; i.e., the 2011–2012 season and the 2013–2014 season. (The 2012–2013 season was shortened due to a lockout).

5.1   Top Pairings

We first present the top pairings according to the impact metrics, as calculated over the entire seasons, for three categories of pairings: forward pairs, defense pairs, and mixed pairs (consisting of a forward and a defender). Tables 3 and 4 summarize these results. Here, we include the player position (defender (D) or the three forward positions: rightwing (R), leftwing (L), and center (C)) and their high-level stats over the entire season (goals (G), assists (A), and plus-minus (+/−)), together with the team, the pairs' total impact (rounded), and the pairs' joint time on ice (TOI), measured in seconds.

Looking first at the 2011–2012 result, we note that many of the names on this list placed high in the scoring race (e.g., Stamkos 2nd, Spezza 4th, Kovalchuk 5th) or were responsible for a large fraction of their teams scoring (e.g., Pavelski/Thornton and O'Reilly/Landeskog). They also all were among the most relied pairings and hence accumulated among the most ice time among all forward pairs. For example, Kovalchuk/Parise, Pavelski/Thornton,

Table 3. Top pairs 2011–2012 according to total impact.

	Player 1				Player 2					Pair stats		
	Name	Pos	G	A	+/-	Name	Pos	G	A	+/-	Team	Impact	TOI
Forwards	Ilya Kovalchuk	R	37	46	-9	Zach Parise	L	31	38	-5	NJD	121.17	40,163
	Ryan O'Reilly	C	18	37	-1	Gabriel Landeskog	L	22	30	+20	COL	115.74	39,021
	Joe Pavelski	C	31	30	+18	Joe Thornton	C	18	59	+17	SJS	112.65	39,353
	Steven Stamkos	C	60	37	+7	Martin St. Louis	R	25	49	-3	TBL	111.77	35,941
	Milan Michalek	L	35	25	+4	Jason Spezza	C	34	50	+11	OTT	111.73	36,689
Defenders	Dan Girardi	D	5	24	+13	Ryan McDonagh	D	7	25	+25	NYR	155.28	55,911
	Filip Kuba	D	6	26	+26	Erik Karlsson	D	19	59	+16	OTT	134.74	47,985
	Francois Beauchemin	D	8	14	-14	Cam Fowler	D	5	24	-28	ANA	125.54	45,795
	Josh Gorges	D	2	14	+14	P.K. Subban	D	7	29	+9	MTL	125.16	44,390
	Carl Gunnarsson	D	4	15	-9	Dion Phaneuf	D	12	32	-10	TOR	123.06	36,181
Mixed	Jason Spezza	C	34	50	+11	Erik Karlsson	D	19	59	+16	OTT	110.58	35,990
	Joe Pavelski	C	31	30	+18	Dan Boyle	D	9	39	+10	SJS	106.04	35,612
	Joe Thornton	C	18	59	+17	Dan Boyle	D	9	39	+10	SJS	102.96	35,160
	Tomas Fleischmann	L	27	34	-7	Brian Campbell	D	4	49	-9	FLA	98.08	31,804
	Stephen Weiss	C	20	27	+5	Brian Campbell	D	4	49	-9	FLA	96.79	32,995

Table 4. Top pairs 2013–2014 according to total impact.

	Player 1				Player 2					Pair stats		
	Name	Pos	G	A	+/-	Name	Pos	G	A	+/-	Team	Impact	TOI
Forwards	James van Riemsdyk	L	30	31	-9	Phil Kessel	R	27	43	-5	TOR	166.42	51,910
	Alex Ovechkin	L	21	38	-35	Nicklas Backstrom	C	18	61	-20	WSH	113.85	40,815
	James van Riemsdyk	L	30	31	-9	Tyler Bozak	C	19	30	+2	TOR	111.39	32,567
	Phil Kessel	R	27	43	-5	Tyler Bozak	C	19	30	+2	TOR	109.57	34,648
	Chris Kunitz	L	35	33	+25	Sidney Crosby	C	36	68	+18	PIT	107.04	36,296
Defenders	Duncan Keith	D	6	55	+22	Brent Seabrook	D	7	34	23	CHI	143.19	46,762
	Dan Girardi	D	5	19	+6	Ryan McDonagh	D	14	29	+11	NYR	129.03	50,102
	Carl Gunnarsson	D	3	14	+12	Dion Phaneuf	D	8	23	31	TOR	124.39	41,219
	Justin Faulk	D	5	27	-9	Andrej Sekera	D	11	33	+4	CAR	121.08	42,966
	Jan Hejda	D	6	11	+8	Erik Johnson	D	9	30	+5	COL	116.57	40,941
Mixed	Phil Kessel	R	27	43	-5	Dion Phaneuf	D	8	23	31	TOR	106.39	32,608
	James van Riemsdyk	L	30	31	-9	Dion Phaneuf	D	8	23	31	TOR	100.82	30,847
	Jason Spezza	C	23	43	-26	Erik Karlsson	D	20	54	-15	OTT	92.91	31,898
	Sean Couturier	C	13	26	+1	Braydon Coburn	D	5	12	-6	PHI	90.43	26,893
	Anze Kopitar	C	29	41	+34	Drew Doughty	D	10	27	+17	LAK	88.74	33,209

and O'Reilly/Landeskog all placed in the top-5 in joint ice time among forward pairs. For the defense pairs and mixed pairs, the correlation was even greater, with four out of five in the corresponding top-five TOI sets.

The 2013–2014 results are similar in that the top-5 list of forward pairs include three of the top-ten names in the scoring race (e.g., Crosby 1st, Kessel 6th, and Ovechkin 8th) and responsible for a large portion of the points on their respective teams, including the line consisting of the three Toronto (TOR) players van Riemsdyk, Kessel, and Bozak. Despite this, none of these three players are still with Toronto, as Kessel was traded to Pittsburgh (PIT) in 2015 and van Riemsdyk and Bozak signed with Philiadelphia (PHI) and St. Louis (STL),

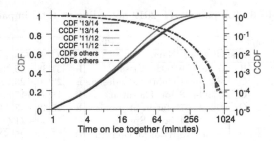

Fig. 1. CDFs and CCDFs of the shared TOI for different NHL seasons.

respectively, the first week of July 2018, illustrating how quickly a team can change direction and build. The other pairings on this list have since combined for three Stanley Cups (PIT in 2015–2016 and 2016–2017 and WSH in 2017–2018), with each of these players contributing to the championships. In fact, Kessel (TOR above) was part of both Pittsburgh (PIT) championship teams. Interestingly, on a related note, the top defense pairings above won the Stanley Cup the 2013–2014 season, as part of the Chicago (CHI) championship team. Keith also won the Norris trophy, as the top defensemen in the league. Other recent Norris trophy winners show up in the top-5 list of the mixed category, including Karlsson (2011–2012 and 2014–2015) and Doughty (2015–2016).

5.2   TOI-Based Analysis

Coaches typically carefully select player combinations and match these against the opponents' lineups so to maximize the chance of success. This means that good players and player combinations typically get more ice time, but also that they may be matched up against tougher competition. We next look at the relationship between TOI and the impact per minute played together for each pair with at least one minute played together during at least one game of the season.

To help interpret the observed relationships, we first present Fig. 1. Here, we plot the cumulative distribution functions (CDFs) and complimentary CDFs (CCDFs) for the joint TOI across all player pairs meeting our threshold criteria, for each of the seasons 2007–2008 through 2013–2014, with the y-axes plotted on linear and logarithmic scales, respectively. First, note that the s-shaped CDFs (plotted on lin-log scale) display close to straight-line behavior on lin-log scale, suggesting that the distribution only has slightly heavier tail than the exponential distribution (for which we would expect straight line behavior) [9]. Second, we note that with exception of the 2012–2013 season (which only had 48 games per team, rather than the typical 82 games per team, due to a player strike/league lockout), the distributions are relatively overlapping, suggesting that this result may be invariant across seasons.

Now, let us get back to the relationship between impact per minute and TOI. Figure 2(a) plots distribution statistics for the impact per minute observed across

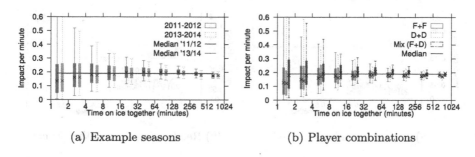

(a) Example seasons   (b) Player combinations

Fig. 2. Impact per minute played together, as a function of the joint TOI.

all player pairs with a joint TOI falling into the time interval $[2^i, 2^{i+1})$, during the 2011–2012 and 2013–2014 seasons, where we vary i from 0 to 9. Here, the body of the bars shows the 75%-iles, the whisker lines the 90%-iles, and the markers ($\times$) the medians. We also include the (almost overlapping) overall medians across all pairs (0.192 and 0.190, respectively). As expected, the variation decreases the more time that pairs play together. However, it is interesting to note that the medians are highest for the pairs that play 16–256 min together, and actually decrease for some of the top TOI pairings. This may partially be due to some pairings (especially on weaker teams) having to be relied upon more than perhaps is healthy for their performance. However, it is also an indication that coaches rely on some of these pairings to play "tough minutes", against the other teams' top players. Yet, the relatively flat median values and overall higher values for pairings that play significant minutes together suggest that coaches overall do a good job distributing the load and/or that more short-term pairings (due to overlapping shifts or bad line changes, where some players are caught out on the ice tired, for example) may perform worse.

When breaking down the analysis based on the player positions of the players in each pair, we have only observed small variations across the traditional combination types: forward pairs and defense pairs. For example, Fig. 2(b) shows relatively similar normalized impact (per minute) for defense pairs and forward pairs, with similar joint TOI. Compared with these categories, the forward-defense pairs have higher impact per pair and contribute with the more pairings (e.g., 7,586 pairs during 2013/14, compared to 4,743 forward and 1,316 defense pairs for the same season). While part of the high-scoring pairings can be attributed to good forwards (e.g., Spezza and Kopitar) being matched with good puck moving defensemen (e.g., Karlsson and Doughty), further analysis of what makes good forward-defense pairings leaves room for interesting future work.

5.3 Relative Ice Time Together

We note that coaches also observe players during practice and off the ice, where part of the chemistry between two players may be developed. It is therefore interesting to analyze if players that spend most of a game together in fact

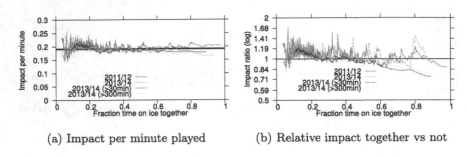

(a) Impact per minute played   (b) Relative impact together vs not

Fig. 3. Impact metrics as a function of fraction of time played together.

produce better during the time they play together or when they play with other players. As a first-cut analysis to look into this question, we plot the impact per minute as function of the time the players play together during the games that they played at least one minute together (Fig. 3(a)) as well as the ratio between the impact per minute when playing together during those games and when not playing together during those same games (Fig. 3(b)). Here, we use an exponentially moving weighted average (EWMA) with $\alpha = 0.02$ to smooth out the curves. Despite using significant smoothing, the variations are significant compared to the relative trends, making it difficult to identify clear patterns. However, in general, it is interesting to see that the players that spend the largest fraction together often have lower relative impact when playing together. At first this may appear counterintuitive. However, for players that also have significant playing time in those games (e.g., subset of the players with at least 300 min together), this may be due to matchups against the other teams' top lines. In other cases, this may be an effect of fourth line players taking their opportunities when on the ice with top line players, with whom they spend less of their total TOI with.

6   Conclusion

In this paper we extend a recent analysis approach for evaluating the performance of players [11] to the related problem of evaluating the performance of pairs of players. In particular, we defined measures for player pairs' impact and analyzed NHL play-by-play data from the 2007–2008 through 2013–2014 NHL seasons using these new metrics.

Our analysis helps identify pairings that have particularly good "chemistry", that performed well across the season (e.g., top pairings), or that the coaches for other reasons rely more heavily on (e.g., that may have played long and tough minutes against other teams' top players). Some of the lessons learned are that for the top pairings, according to the impact metrics, many of the names on this list placed high in the scoring race or were responsible for a large fraction of their teams scoring. Further, forward-defense pairs have higher impact per pair, and the players that spend the largest fraction together often have lower relative impact when playing together. Using the data, we also hypothesize that coaches desire to play their top players against the other teams' top players appears to even out the relative impact per minute observed across different player categories.

Regarding future work, one direction is to work with different reward functions in the Q-learning algorithm to investigate impact of player actions for different desirable outcomes (e.g., shots on goals, powerplays). We also intend to investigate alternative pair impact definitions. For instance, the current definition credits the pair for the actions when they are on the ice (indirect impact), while it would be interesting to compare to direct impact as well (e.g., the pair receives credit only for direct impact actions by one of the players in the pair).

References

1. 82games: Which players play well together? - 2003–04 season. http://www.82games.com/ppairs0304.htm
2. 82games: Which players play well together? - 2004–05 season. http://www.82games.com/ppairs0405.htm
3. Ayer, R.: Big 2's and Big 3's: analyzing how a team's best players complement each other. In: MIT Sloan Sports Analytics Conference (2012)
4. Gramacy, R.B., Jensen, S.T., Taddy, M.: Estimating player contribution in hockey with regularized logistic regression. J. Quant. Anal. Sports **9**, 97–111 (2013). https://doi.org/10.1515/jqas-2012-0001
5. Kaplan, E.H., Mongeon, K., Ryan, J.T.: A Markov model for hockey: manpower differential and win probability added. INFOR Inf. Syst. Oper. Res. **52**(2), 39–50 (2014). https://doi.org/10.3138/infor.52.2.39
6. Liu, G., Schulte, O.: Deep reinforcement learning in ice hockey for context-aware player evaluation. In: Lang, J. (ed.) Proceedings of the Twenty-Seventh International Joint Conference on Artificial Intelligence, pp. 3442–3448 (2018). https://doi.org/10.24963/ijcai.2018/478
7. Macdonald, B.: A regression-based adjusted plus-minus statistic for NHL players. J. Quant. Anal. Sports **7**(3) (2011). https://doi.org/10.2202/1559-0410.1284
8. Macdonald, B.: An improved adjusted plus-minus statistic for NHL players. In: MIT Sloan Sports Analytics Conference (2011)
9. Mahanti, A., Carlsson, N., Mahanti, A., Arlitt, M., Williamson, C.: A tale of the tails: power-laws in internet measurements. IEEE Netw. **27**(1), 59–64 (2013). https://doi.org/10.1109/MNET.2013.6423193
10. Pettigrew, S.: Assessing the offensive productivity of NHL players using in-game win probabilities. In: MIT Sloan Sports Analytics Conference (2015)
11. Routley, K., Schulte, O.: A Markov game model for valuing player actions in ice hockey. In: Meila, M., Heskes, T. (eds.) Uncertainty in Artificial Intelligence, pp. 782–791 (2015)
12. Schuckers, M., Curro, J.: Total Hockey Rating (THoR): a comprehensive statistical rating of National Hockey League forwards and defensemen based upon all on-ice events. In: MIT Sloan Sports Analytics Conference (2013)
13. Schulte, O., Khademi, M., Gholami, S., Zhao, Z., Javan, M., Desaulniers, P.: A Markov game model for valuing actions, locations, and team performance in ice hockey. Data Min. Knowl. Discov. **31**(6), 1735–1757 (2017). https://doi.org/10.1007/s10618-017-0496-z

14. Schulte, O., Zhao, Z., Javan, M., Desaulniers, P.: Apples-to-apples: clustering and Ranking NHL players using location information and scoring impact. In: MIT Sloan Sports Analytics Conference (2017)
15. Thomas, A., Ventura, S.L., Jensen, S., Ma, S.: Competing process hazard function models for player ratings in ice hockey. Ann. Appl. Stat. **7**(3), 1497–1524 (2013)

Model Trees for Identifying Exceptional Players in the NHL and NBA Drafts

Yejia Liu[✉], Oliver Schulte, and Chao Li

Department of Computing Science, Simon Fraser University, Burnaby, Canada
{yejial,oschulte,chao_li_2}@sfu.ca

Abstract. Drafting players is crucial for a team's success. We describe a data-driven interpretable approach for assessing prospects in the National Hockey League and National Basketball Association. Previous approaches have built a predictive model based on player features, or derived performance predictions from comparable players. Our work develops model tree learning, which incorporates strengths of both model-based and cohort-based approaches. A model tree partitions the feature space according to the values or learned thresholds of features. Each leaf node in the tree defines a group of players, with its own regression model. Compared to a single model, the model tree forms an ensemble that increases predictive power. Compared to cohort-based approaches, the groups of comparables are discovered from the data, without requiring a similarity metric. The model tree shows better predictive performance than the actual draft order from teams' decisions. It can also be used to highlight the strongest points of players.

Keywords: Player ranking · Logistic Model Trees ·
M5 regression trees · National Hockey League ·
National Basketball Association

1 Introduction

Player ranking is one of the most studied subjects in sports analytics [1]. In this paper we consider predicting success in both the National Hockey League (NHL) and National Basketball Association (NBA) from pre-draft data, with the goal of supporting draft decisions. The publicly available pre-draft data aggregate a season's performance into a single set of numbers for each player. Our method can be applied to any data of this type, for example also to soccer draft data. Since our goal is to support draft decisions by teams, we ensure that the results of our data analysis method can be easily explained to and interpreted by sports experts. Previous approaches for analyzing NHL/NBA draft data take a regression approach or a similarity-based approach [7,19]. Regression approaches build a predictive model that takes as input a set of player features, such as demographics (age, height, weight) and pre-draft performance metrics (goals scored, minutes played), and output a predicted *success metric* (e.g. number

© Springer Nature Switzerland AG 2019
U. Brefeld et al. (Eds.): MLSA 2018, LNAI 11330, pp. 93–105, 2019.
https://doi.org/10.1007/978-3-030-17274-9_8

of games played, or player efficiency rating). Cohort-based approaches divide players into groups of comparables and predict future success based on a player's cohort. For example, the PCS model [28] clusters players according to age, height, and scoring rates. One advantage of the cohort model is that predictions can be explained by reference to similar known players, which many domain experts find intuitive. For this reason, several commercial sports analytics systems, such as Sony's Hawk-Eye system, have been developed to identify groups of comparables for each player. Our aim in this paper is to describe a new model for draft data that combines the advantages of both approaches, regression-based and similarity-based. Our method uses a model tree [4,13]. Each node in the tree defines a new yes/no question, until a leaf is reached. Depending on the answers to the questions, each player is assigned a group corresponding to a leaf. The tree builds a different regression model for each leaf node. Figure 1 shows an example model tree. A model tree offers several advantages.

- Compared to a single regression model, the tree defines *an ensemble of regression models*, based on *non-linear thresholds*. This increases the expressive power and predictive accuracy of the model. The tree can represent complex interactions between player features and player groups.
- Compare to a similarity-based model, *tree construction learns groups of players from the data*, without requiring the analyst to specify a similarity metric. Because tree learning selects splits that increase predictive accuracy, the learned distinctions between the groups are guaranteed to be predictively relevant to a player's future career success. Also, the tree creates a model, not a single prediction, for each group, which allows it to differentiate players from the same group.

In the NHL draft, we approach prospect ranking not by directly predicting the future number of games played, but by *predicting whether a prospect will play any number of games at all in the NHL*, due to an excess-zeros problem [11]. For the NBA draft, we use a linear regression model tree that directly predicts the continuous success variable.

Following the work of Schuckers et al. and Greene [7,19], we evaluate the model trees ranking results by comparing to a ranking based on the players' actual future success, measured as the number of career games they played after 7 years in the NHL, or player efficiency rating (PER) in the NBA. The rank correlation of our logistic regression ranking for the NHL draft is competitive with that achieved by the generalized additive model of [19], which is currently the state-of-the-art for NHL draft data. As for the NBA draft, our linear regression ranking performs better than the actual draft pick in terms of correlations with players' future success.

We show in case studies that the feature weights learned from the data can be used to *explain the ranking* in terms of which player features contribute the most to an above-average ranking. In this way, the model tree can be used to highlight exceptional features of a player that scouts and teams can take into account in their evaluation.

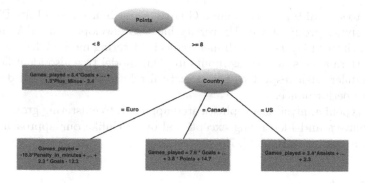

Fig. 1. A simple model tree example.

2   Related Work

Different approaches to player ranking are appropriate for different data types. For example, with dynamic play-by-play data, Markov models have been used to rank players [2, 20, 24]. For data that record the presence of players when a goal is scored, regression models have also been applied to extend the classic wins-contribution metrics [15, 22]. In this paper, we utilize player statistics that aggregate a season's performance into a single set of numbers. While these data are much less informative than play-by-play data, they are easier to obtain, interpret, and process.

Regression Approaches. To our knowledge, this is the first application of model trees to hockey draft prediction, and the first model for predicting whether a draftee plays any games at all. The closest predecessor to our work is due to Schuckers [19] and Greene [7], who use a single linear regression model to predict future success from junior league data.

Similarity-Based Approaches assume a similarity metric and group similar players to predict performance. A sophisticated example from baseball is the nearest neighbour analysis in the PECOTA system [23]. For ice hockey, the Prospect Cohort Success (PCS) model [28], cohorts of draftees are defined based on age, height, and scoring rates. Model tree learning provides an automatic method for identifying cohorts with predictive validity. We refer to cohorts as groups to avoid confusion with the PCS concept. Because tree learning is computationally efficient, our model tree is able to take into account a larger set of features than age, height, and scoring rates. Also, it provides a separate predictive model for each group that assigns group-specific weights to different features. In contrast, PCS makes the same prediction for all players in the same cohort. So far, PCS has been applied to predict whether a player will score more than 200 games career total. Tree learning can easily be modified to make predictions for any game count threshold. For basketball, many clustering approaches focus on defining appropriate roles or positions for a player. In Lutz's work [14], NBA players are clustered to several types like Combo Guards, Floor Spacers and Elite Bigs. The sports analytics group of Yale University has also developed an NBA clustering system to cluster players through hierarchical clustering methodology with their season performance statistics as inputs [6]. Our model tree also identifies which positions differ with respect to the statistical relationships between draft data and future performance.

Archetypoid analysis is a sophisticated approach to clustering groups of comparable players and identifying exceptional ones. Unlike our approach, it does not build a predictive model [27].

3   Datasets

We describe our dataset sources and preprocessing steps.

Ice Hockey Data. Our data were obtained from public-domain on-line sources, including nhl.com, eliteprospects.com, and thedraftanalyst.com. We are also indebted to David Wilson for sharing his NHL performance dataset [29]. The full dataset is posted on the worldwide web[1]. We consider players drafted into the NHL between 1998 and 2008 (excluding goaies) and divided them into two cohorts (1998–2002 cohort and 2004–2008 cohort). Our data include demographic factors (e.g. age, weight, height), performance metrics for the year in which a player was drafted (e.g., goals scored, plus/minus), career statistics (e.g. number of games played, time on ice), and the rank assigned to a player by the NHL Central Scouting Service (CSS). If a player was not ranked by the CSS, we assigned 1+ the maximum rank for his draft year to his CSS rank value. Another preprocessing step was to pool all European countries into a single category. If a player played for more than one team in his draft year (e.g., a league team and a national team), we added up the counts from different teams. We also eliminated players drafted in 2003 since most of them have no CSS rank [19].

Basketball Data. Our basketball datasets were obtained from https://www.basketball-reference.com, a rich resource of NBA player data, containing both pre-draft and career information. We posted our data in the worldwide web[2]. We considered players drafted into NBA between 1985 and 2011. Our training data included statistics of players drafted from 1985 to 2005, while players drafted from 2006 to 2011 were considered as our testing data. We excluded players whose college performance statistics are not available. For the 15 drafted players whose career statistics are not available, we replaced their career PER by $min(x) - std(x)$, where $min(x)$ is the minimum career PER and $std(x)$ is the standard deviation of career PER for all players drafted in the same year. The motivation is that these players are very likely judged to be worse, by coaches and team experts, than players who played in the NBA. We leave other imputation methods for future work.

---

[1] https://github.com/liuyejia/Model_Trees_Full_Dataset.
[2] https://github.com/sfu-cl-lab/Yeti-Thesis-Project/tree/master/NBA_work.

4   Methodology

4.1   Success Metrics

In the NHL draft, we took as our dependent variable *the total number of games* g_i *played by a player i after 7 years* under an NHL contract. The first seven seasons are chosen because NHL teams have at least seven year rights to players after they are drafted [29]. Other interesting success metrics like the Wins Above Replacement (WAR) or Total Hockey Rating (ThoR) usually require play-by-play data and are often computationally expensive [18,25]. For the NBA, we adopted *player efficiency rating (PER)* as our success metric, which encompasses a large set of inputs and takes nearly every aspect of a player's contribution into consideration. The calculation is as follows:

$$PER = (uPER \times \frac{lg\,Pace}{tm\,Pace}) \times \frac{15}{lg\,uPER}$$

where $uPER$ is the unadjusted PER, calculated using player performance variables, as well as team and league statistics (e.g. game pace). In the above formula, lg is indicates the league, tm the team [9]. PER is widely used, but other metrics have been proposed [21]. Model trees can be applied with any performance metric.

4.2   Model Trees

Model Trees are a flexible formalism that can be built for any regression model. We use a logistic regression model for the NHL and linear regression for the NBA.

Logistic Model Tree (NHL). Our logistic regression model tree *predicts whether a player will play any games at all in the NHL* ($g_i > 0$). The motivation is that many players in the draft never play any NHL games at all (up to 50% depending on the draft year) [26]. This poses an extreme *excess-zeros problem* for predicting directly the number of games played. In contrast, for the classification problem of predicting whether a player will play any NHL games, excess-zeros means that the dataset is balanced between the classes. This classification problem is interesting in itself; for instance, a player agent would be keen to know what chances their client has to participate in the NHL. The logistic regression probabilities $p_i = P(g_i > 0)$ can be used not only to predict whether a player will play any NHL games, but also to rank players such that the ranking correlates well with the actual number of games played. We built a model tree whose leaves contain a logistic regression model, where each player can be assigned to a unique leaf node to compute a probability p_i.

Figure 2 shows the logistic regression model tree learned for our second cohort (players drafted between 2004–2008). It places *CSS rank* at the root as the most important attribute. Players ranked better than 12 form an elite group, of whom almost 82% play at least one NHL games. For players at rank 12 or below, the

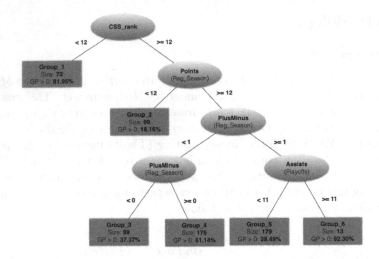

Fig. 2. Logistic Regression Model Trees (LMT) for the 2004, 2005, 2006 cohort in NHL. The tree was built using the LogitBoost algorithm implemented in the LMT package of the Weka Program [3,8]. Each leaf defines a group of players. For each group, the figure shows the proportion of players who played at least one game in the NHL. Each leaf contains a logistic regression model for its group (not shown), which produces different predictions for players from the same group but with different features. CSS_rank denotes rankings from the Central Scouting Service; lower rank numbers are better (e.g. rank 1 is the best).

tree considers next their regular season points total. Players with rank and total *points* below 12 form an unpromising group: only 16% of them play an NHL game. Players with rank below 12 but whose points total is 12 or higher, are divided by the tree into three groups according to whether their *regular season plus/minus* score is positive, negative, or 0. (A three-way split is represented by two binary splits). If the plus/minus score is negative, the prospects of playing an NHL game are fairly low at about 37%. For a neutral plus/minus score, this increases to 61%. For players with a positive plus/minus score, the tree uses the number of *playoffs assists* as the next attribute. Players with a positive plus/minus score and more than 10 playoffs assists form a small but strong group that is 92% likely to play at least one NHL game.

Linear Regression Tree (NBA). Different from NHL, most drafted basketball players played at least one game in the NBA (over 80% in our datasets depending on the draft year). Since there is no excess-zeros issue, we used linear regression, which links predictors to the continuous dependent variable directly. Similar to the construction of logistic model tree in the NHL draft, we built a linear regression tree whose leaves contain a linear regression model. Each player can be assigned to its own leaf node to predict his future career PER.

Figure 3 shows our learned M5 regression model tree [16]. The root attribute is *position*, as the most important attribute due to its highest information gain.

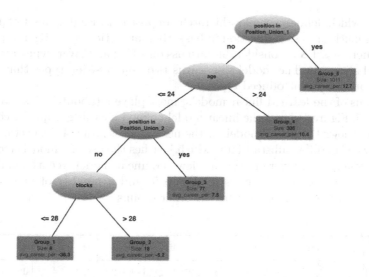

Fig. 3. M5 regression trees for the drafted players in 1985–2011 drafts. The tree was built with the M5P package of Weka [3,8]. Each leaf defines a group of players, and a linear regression model for its group (not shown). For each group, the figure shows the average career PER. The group model assigns weights to *all* player features, and produces different predictions for players from the same group with different features. The values of Position_Union_1 and Position_Union_2 are listed in [12].

Players who are from Position_Union_1 have an average PER about 13, forming a better group compared to other players. Position_Union_1 is a list of player positions (e.g. Center), automatically grouped by the M5P algorithm. For players who are not from Position_Union_1, the tree takes *age* as the next splitting attribute. Players who are older than 24 years old and are not from Position_Union_1, belong to a less promising group with PER around 10. Then, the tree chooses *position* as another splitting point again, reflecting its significance. For players who do not belong to Position_Union_1 but belong to Position_Union_2, with age smaller than or equal to 24, they form an average level group. The average PER value of players in this group is around 7. Position_Union_2 is another list of positions, which were automatically grouped by the M5P algorithm. Lastly, the tree chooses *blk (blocks)* as the splitting feature. The sizes of Group 1 and Group 2 are relatively small (8 and 18). These are special groups that require a customized model, according to tree. The tree finds conditions that separate the strong group 5 from the groups 1 and 2 that show substantially worse average performance.

For discrete variables like Position, the M5P algorithm has the capacity to evaluate splits based not only on specific values (e.g. Position = Center), but also on a disjunction of values (e.g. Position = Center or Forward). The disjunctions of positions represented by Position_Union_1 and Position_Union_2 offered the best trade-off between model complexity and data fit. Moreover, in its leaf model, the tree groups the positions further into subgroups (3 for Group 5 and 4 for

Group 4), which defines a shallow hierarchy of positions. Finding a set of position types is a much discussed question in basketball analytics [5,14,21] into position types. There has been considerable discussion of what player types are useful in basketball [6,14]. The model tree learns two new position types that are the union of previously introduced types.

In terms of the learned linear models, most players (about 72% are assigned to Group 5. Figure 4 shows the linear model defined for this group. For contrast, the Figure shows the linear model for the next-biggest group 4. Inspection shows that the weights differ substantially, which justifies the use of a model tree rather than a single model. Overall, we can view the linear model tree as defining one standard model for most players in Group 5, and a number of more specific models for special cases defined by the other groups.

Metrics / Group	age	position	g	mp	ft	fta	trb	ast	blk	pts	ah
4	-34.35	1.92 0.05 0.07	12.89	all: 0 per: 5.08	all: 2.25 per: -0.14	-18.63	all: 0.21 per: 0	all: 0 per: 0.11	9.51	all: -11.24 per: 17.91	0.04
5	-2.39	0.36 0.83 0.53 1.34	-6.54	all: 4.27 per: -4.19	all: 10.57 per: -5.33	-10.69	all: 10.05 per: -4.95	all: 5.66 per: 0.04	2.41	all: 2.92 per: 4.01	1.03

Fig. 4. Example of learned weights. The table shows the weights of each feature in the linear equation for Group 4 and Group 5. *Per* represents per game statistics and *all* denotes overall statistics. g = games played, mp = minutes played, ft = free throws, fta = free throws attempt, trb = total rebounds, ast = assists, blk = blocks, pts = points, ah = amature honor. For possible values of Position, see the text.

5   Results

5.1   Modelling Results of the NHL Draft

Following [19], we consider three rankings as follows:

1. The performance ranking based on the actual number of NHL games that a player played.
2. The ranking of players based on the probability p_i of playing at least one game (Tree Model SRC).
3. The ranking of players based on the order in which they were drafted by team (Draft Order SRC).

Table 1 gives results for the out of sample prediction for players drafted in each of the four out of sample drafts using *games played* (GP) as the response variable. The draft order can be viewed as the ranking that reflects the judgment of NHL teams. Like [19], we evaluate the predictive accuracy of the Draft Order and the LMT model using the Spearman Rank Correlation (SRC) between (i)

Draft order and actual number of games, and (ii) p_i order and actual number of games, as shown in Table 1. We can see from Table 1 that the average rank correlation between the NHL draft order and the NHL Performance metric is about 0.43 while our tree model averages about 0.81 for both cohorts. This strongly suggests that our model outperforms the draft ordering by player selection.

Table 1. Predictive Performance (Draft Order, our Logistic Model Trees) measured by Spearman Rank Correlation (SRC) with actual number of games. Bold indicates the best values.

Training data NHL draft years	Out of sample draft years	Draft Order SRC	LMT classification accuracy	LMT SRC
1998, 1999, 2000	2001	0.43	82.27%	**0.83**
1998, 1999, 2000	2002	0.30	85.79%	**0.85**
2004, 2005, 2006	2007	0.46	81.23%	**0.84**
2004, 2005, 2006	2008	0.51	63.56%	**0.71**

5.2 Modelling Results of the NBA Draft

We used both the Pearson Correlation and the Spearman Rank Correlation to compare the predictive power of our tree models to the actual draft order and a baseline method (Ordinary Linear Regression). As shown in Table 2, our model tree performs better than the actual draft order and ordinary linear regression.

Table 2. Comparison of predictive performance between draft order, linear regression and our tree models. Bold indicates the best values.

Method	Evaluation		
	Pearson Correlation	Spearman Rank Correlation	RMSE
Draft order	0.42	0.39	NaN
Ordinary linear regression	0.45	0.40	7.14
Our Model Tree	**0.55**	**0.43**	**6.16**

6 Case Studies: Exceptional Players and Their Strong Points

Teams make drafting decisions not based on player statistics alone, but drawing on all relevant source of information, and with extensive input from scouts and other experts. As Cameron Lawrence from the Florida Panthers put it, "the numbers are often just the start of the discussion" [10]. In this section we discuss how the model tree can be applied to support the discussion of individual players by highlighting their special strengths. The idea is that the learned weights can be used to identify which features of a highly-ranked player differentiate him the most from others in his group.

6.1 Explaining the Rankings: Identify Strong Points

Our method is as follows. For each group, we find the average feature vector of the players in the group, which we denote by $(\overline{x_{g_1}}, \overline{x_{g_2}}, ..., \overline{x_{g_m}})$. We denote the features of player i as $(x_{i_1}, x_{i_2}, ..., x_{i_m})$. Then given a weight vector $(w_1, ..., w_m)$ for the logistic regression model of group g, the log-odds difference between player i and a random player in the group is given by

$$\sum_{j=1}^{m} w_j(x_{i_j} - \overline{x_{g_i}})$$

We can interpret this sum as a measure of how high the model ranks player i compared to other players in his group. This suggests defining as the player's strongest features the x_{i_j} that maximize $w_j(x_{i_j} - \overline{x_{g_i}})$. This approach highlights features that are (i) relevant to predicting future success, as measured by the magnitude of w_j, and (ii) different from the average value in the player's group of comparables, as measured by the magnitude of $x_{i_j} - \overline{x_{g_i}}$. Table 3 summarizes the strongest players and their strongest statistics in the NHL and NBA draft.

Table 3. Strongest statistics for the top three players in the strongest group for the NHL and NBA draft [4].

Top players	Strongest points ($\overline{x}$ = group mean)		
NIIL			
Sidney Crosby	Points *(regular season)* 188 ($\overline{x} = 47$)	Assists *(regular season)* 110 ($\overline{x} = 27$)	CSS_rank 1 ($\overline{x} = 7$)
Patrick Kane	Points *(regular season)* 154 ($\overline{x} = 47$)	Assists *(regular season)* 87 ($\overline{x} = 27$)	CSS_rank 2 ($\overline{x} = 7$)
Sam Gagner	Points *(regular season)* 118 ($\overline{x} = 47$)	Assists *(playoffs)* 22 ($\overline{x} = 4$)	Assists *(regular season)* 83 ($\overline{x} = 27$)
NBA			
Larry Johnson	Free throws 162 ($\overline{x} = 122$)	Total rebounds 380 ($\overline{x} = 214$)	Assists 104 ($\overline{x} = 84$)
Anfernee Hardaway	Total rebounds 273 ($\overline{x} = 214$)	Assists 204 ($\overline{x} = 84$)	Minutes played 1196 ($\overline{x} = 929$)
Chris Webber	Total rebounds 362 ($\overline{x} = 84$)	Minutes played 1143 ($\overline{x} = 929$)	Assists 90 ($\overline{x} = 84$)

6.2   Case Studies

NHL Draft. *Sidney Crosby* and *Patrick Kane* are obvious stars, who have outstanding statistics even relative to other players in this strong group. We see that the ranking for individual players is based on different features, even within the same group. The table also illustrates how the model allows us to identify a group of comparables for a given player.

The most interesting cases are often those whose ranking differs from the scouts' CSS rank. Among the players who were not ranked by CSS at all, our model ranks *Kyle Cumiskey* at the top. Cumiskey was drafted in place 222, played 132 NHL games in his first 7 years, represented Canada in the World Championship, and won a Stanley Cup in 2015 with the Blackhawks. His *strongest points* were being Canadian, and the number of games played (e.g., 27 playoffs games vs. 19 group average). In the lowest CSS-rank group, Group 6, our top-ranked player *Brad Marchand* received CSS rank 80, even below his Boston Bruin teammates Lucic's. Given his Stanley Cup win and success representing Canada, arguably our model was correct to identify him as a strong NHL prospect. The model *highlights* his superior play-off performance, both in terms of games played and points scored.

NBA Draft. The most prestigious player *Chris Webber* identified by our model was a superstar in the NBA. He is a five-time NBA All-Star, a five-time All-NBA Team member, and NBA Rookie of the Year. His *strongest points* in his pre-draft year are trb (total rebounds), mp (minutes played) and ast (assists). There are also examples of players highlighted by our model, who in hindsight were undervalued compared to other players from the same cohort. For instance, *Matt Geiger* was ranked at 13*th* in his group by our model tree, who was picked at 42*th* in the draft, after Todd Day (8th) and Bryant Stith (13th). However, his career PER is 15.2, above these two players drafted before him. His pre-draft *strongest points* were identified as total rebounds, assists and minutes played. A more recent instance is *Dejuan Blair*, who had the 37*th* overall draft pick in 2009, taken after Jordan Hill (8th), Ricky Rubio (5th), but he obtained almost the same career PER as them.

7   Conclusion

We have proposed building regression model trees for ranking draftees in the NHL and NBA, or other sports, based on a list of player features and performance statistics. The model tree groups players according to the values of discrete features, or learned thresholds for continuous performance statistics. Each leaf node defines a group of players that is assigned its own regression model. Tree models combine the strength of both regression and cohort-based approaches, where player performance is predicted with reference to comparable players.

Key findings include the following: (1) The model tree ranking correlates well with the actual success ranking according to the actual number of games played, better than draft order. (2) The model tree can highlight the exceptionally strong points of draftees that make them stand out compared to the other players in their group.

Tree models are flexible and can be applied to other prediction problems to discover groups of comparable players as well as predictive models. For example, we can predict future NHL success from past NHL success, similar to Wilson [29] who used machine learning models to predict whether a player will play more than 160 games in the NHL after 7 years. Another direction is to apply the model to other sports, for example drafting for the National Football League.

Future Work. A common issue in drafting is comparing players from different leagues. A model tree offers a promising approach to this issue as it can make data-driven decisions about whether players from different leagues should be assigned to different models. For example, preliminary experiments with the NHL data show that adding the junior league of a player as a feature leads, the tree to split on the player's country. This finding suggests that model tree learning can be applied to assess the skills of both North American and international NBA prospects in a single model. Identifying which skills of international players make them compatible with the NBA has become increasingly important over the past decades [17]. Building a model tree to make up-to-date predictions for the current draft would be the most relevant application for teams.

References

1. Albert, J., Glickman, M.E., Swartz, T.B., Koning, R.H.: Handbook of Statistical Methods and Analyses in Sports. CRC Press, Boca Raton (2017)
2. Cervone, D., D'Amour, A., Bornn, L., Goldsberry, K.: POINTWISE: predicting points and valuing decisions in real time with NBA optical tracking data. In: MIT Sloan Sports Analytics Conference (2016)
3. Frank, E., Hall, M., Witten, I.: The WEKA workbench. Online appendix for data mining: practical machine learning tools and techniques (2016)
4. Friedman, J., Hastie, T., Tibshirani, R.: Additive logistic regression: a statistical view of boosting. Ann. Stat. **28**(2), 337–407 (2000)
5. Green, E., Menz, M., Benz, L., Zanuttini-Frank, G., Bogaty, M.: Clustering NBA players (2016). https://sports.sites.yale.edu/clustering-nba-players
6. Green, E., Menz, M., Benz, L., Zanuttini-Frank, G., Bogaty, M.: Clustering NBA players (2017). https://sports.sites.yale.edu/clustering-nba-players
7. Greene, A.C.: The success of NBA draft picks: can college career predict NBA winners. Master's thesis, St. Cloud State University (2015)
8. Hall, M., Frank, E., Homes, G., Pfahringer, B., Reutemann, P., Witten, I.: The WEKA data mining software: an update. SIGKDD Explor. **11**, 10–18 (2009)
9. Hollinger, J.: Calculating PER (2013). https://www.basketball-reference.com/about/per.html
10. Joyce, E., Lawrence, C.: Blending old and new: how the Florida panthers try to predict future performance at the NHL entry draft (2017)
11. Lachenbruch, P.A.: Analysis of data with excess zeros. Stat. Methods Med. Res. **11** (2002)

12. Liu, Y., Schulte, O., Hao, X.: Drafting NBA players based on their college performance (2018). https://github.com/liuyejia/Model_Trees_Full_Dataset/blob/master/NBA_work/ReadMe_NBA.md
13. Loh, W.Y.: GUIDE user manual. University of Wisconsin-Madison (2017)
14. Lutz, D.: A cluster analysis of NBA players. In: MIT Sloan Sports Analytics Conference (2012)
15. Macdonald, B.: An improved adjusted plus-minus statistic for NHL players. In: MIT Sloan Sports Analytics Conference (2011)
16. Quinlan, J.R.: Learning with continuous classes, pp. 343–348. World Scientific (1992)
17. Salador, K.: Forecasting performance of international players in the NBA. In: MIT Sloan Sports Analytics Conference (2011)
18. Schuckers, M., Curro, J.: Total Hockey Rating (THoR): a comprehensive statistical rating of National Hockey League forwards and defensemen based upon all on-ice events. In: MIT Sloan Sports Analytics Conference (2013)
19. Schuckers, M.E., Statistical Sports Consulting, LLC: Draft by numbers: using data and analytics to improve National Hockey League (NHL) player selection. In: MIT Sloan Sports Analytics Conference (2016)
20. Schulte, O., Zhao, Z., SPORTLOGiQ: Apples-to-apples: clustering and ranking NHL players using location information and scoring impact. In: MIT Sloan Sports Analytics Conference (2017)
21. Shea, S., Baker, C.: Basketball Analytics: Objective and Efficient Strategies for Understanding How Teams Win. CRC Press, Boca Raton (2017)
22. Sill, J.: Improved NBA adjusted +/− using regularization and out-of-sample testing. In: MIT Sloan Sports Analytics Conference (2010)
23. Silver, N.: PECOTA 2004: A Look Back and a Look Ahead, pp. 5–10. Workman Publishers, New York (2004)
24. Thomas, A., Ventura, S., Jensen, S., Ma, S.: Competing process hazard function models for player ratings in ice hockey. Ann. Appl. Stat. 7(3), 1497–1524 (2013)
25. Thomas, A.C., Ventura, S.: The highway to WAR: defining and calculating the components for wins above replacement (2015). https://aphockey.files.wordpress.com/2015/04/sam-war-1.pdf
26. Tingling, P., Masri, K., Martell, M.: Does order matter? An empirical analysis of NHL draft decisions. Sport Bus. Manag.: Int. J. 1(2), 155–171 (2011)
27. Vinué, G., Epifanio, I.: Archetypoid analysis for sports analytics. Data Min. Knowl. Discov. 31(6), 1643–1677 (2017)
28. Weissbock, J.: Draft analytics: unveiling the prospect cohort success model. Technical report (2015). https://canucksarmy.com/2015/05/26/draft-analytics-unveiling-the-prospect-cohort-success-model/
29. Wilson, D.R.: Mining NHL draft data and a new value pick chart. Master's thesis, University of Ottawa (2016)

Evaluating NFL Plays: Expected Points Adjusted for Schedule

Konstantinos Pelechrinis[1]([✉]), Wayne Winston[2], Jeff Sagarin[3], and Vic Cabot[2]

[1] University of Pittsburgh, Pittsburgh, USA
kpele@pitt.edu
[2] Indiana University, Bloomington, USA
{winston,NA}@indiana.edu
[3] USA Today, McLean, USA
jsagarin@attglobal.net

Abstract. "Not all yards are created equal". A 3rd and 15, where the running back gains 12 yards is clearly less valuable than a 3rd and 3 where the running back gains 4 yards, even though it will not necessarily show up in the yardage statistics. While this problem has been addressed to some extent with the introduction of expected point models, there is still another inequality omission in the creation of yards and this is the opposing defense. Gaining 6 yards on a 3rd and 5 against the top defense is not the same as gaining 6 yards on a 3rd and 5 against the worst defense. Adjusting these expected points model for opponent strength is thus crucial. In this paper, we develop an optimization framework that allows us to compute offensive and defensive ratings for each NFL team and consequently adjust the expected point values accounting for the opposition faced. Our framework allows for assigning different point values to the offensive and defensive units of the same play, which is the rational thing to do especially in a league with an uneven schedule such as the NFL. The average absolute difference between the raw and adjusted points is 0.07 points/play (p-value < 0.001), while the median discrepancy is 0.06 (p-value < 0.001). This might seem negligible, but with an average of 130 plays per game this translates to approximately 125 points/season discredited. The opponent strength adjustment that we introduce in this work is crucial for obtaining a better evaluation of personnel based on the actual competition they faced. Furthermore, our work allows to evaluate special teams' performance, a unit that has been identified as crucial for winning but has never been properly evaluated. We firmly believe that our developed framework can make significant strides towards computing an accurate estimate of the true monetary worth of a given player.

1 Introduction

Suppose it is a 3rd down and 15 yards for the Cowboys on their own 40-yard line and Ezekiel Elliott gains 12 yards. From the standpoint of Fantasy Football and

V. Cabot—1965–1994.

© Springer Nature Switzerland AG 2019
U. Brefeld et al. (Eds.): MLSA 2018, LNAI 11330, pp. 106–117, 2019.
https://doi.org/10.1007/978-3-030-17274-9_9

yards-per-rushing attempt this is a very good play. In terms of the Cowboys' expected margin of victory in the game, this play is bad because the Cowboys will probably punt to the opposition. What is needed to properly evaluate each NFL play is a measure of the "expected point value" generated by each play. Cincinnati Bengal's QB Virgil Carter and his thesis advisor Machol [5] were the first to realize the importance of assigning a point value to each play. Carter and Machol estimated the value of first and 10 on every 10-yard grouping of yard lines by averaging the points scored on the next scoring event. In *The Hidden Game of Football* [4] Carroll, Palmer and Thorn attempted to determine a linear function that gave the value of a down and yards to go situation. In this work we make use of the expected points model developed by Winston, Cabot and Sagarin [8,9] (WCS for short) that models the game of American football as an infinite horizon zero-sum stochastic game. This model defines the value of the game in its current *state* to be the expected number of points by which the offensive team will beat the defensive team if the current score is tied. The state is defined by the triplet: $<Down, Yards - to - Go, Field\ Position>$. This state definition results in 19,799 possible states. Since there are roughly 40,000 plays in an NFL season most states will occur rarely (if at all!). Examining the play-by-play data for the 2014–2016 seasons we identify that there are only 4,972, 4,989 and 4,966 states appearing in each of the season respectively. Furthermore, among these there is a prominent state (that is, the touchback state) in each season with almost 5x more appearances than the second most frequent state. This makes it challenging to use play-by-play data to accurate evaluate the value of each state, since many of these states do not appear in the data (or they appear very rarely). WCS details are beyond the scope of this work but in brief, WCS computes for each of the 19,799 states the value of the game and the corresponding mixed optimal strategies by using a linear programming approach [7] (Appendix A). For example, WCS provides a value of -2.575 for the state $<4,\ 4,\ 7>$, which means that for a team that has a fourth down, at their own 7-yard line and need 4 more yards for a first down, they will be outscored by approximately 2.6 points from that point on. Using the values of the starting and ending state of a play we can obtain the corresponding play value. For instance, consider a team with the ball 1st and 10 on their own 23, which corresponds to a state value of $v_1 = -0.014$. Their first down play gains 4 yards and hence, the state now is 2nd and 6 on their own 27, which has a value of $v_2 = 0.109$. The value of this play is $v_2 - v_1 = 0.123$. More recently, other expected points models have appeared in the literature. For example, Yurko *et al.* [10] used a multinomial logistic regression to estimate the state values from play-by-play data, while Brian Burke [1] grouped *similar* states together and used the next score (i.e., which team scores next - offense or defense) to obtain an estimate for the current state's value.

Each one of these approaches clearly have its own merits and drawbacks. However, a crucial drawback that all of them exhibit is that none adjusts the value of a play for the strength of opponent. For example, a running play with a value of .10 points against a team with an above average rushing defense should be more valuable than the same play against a team with a below average running defense. Adjusting for the strength of the opponent can provide us with a more holistic view of a play's value. Therefore, the major contribution of our work is determining the points per play for each team's running, passing, and special teams units, **adjusted for schedule strength**[1].

In the following section we present our adjustment optimization framework, while Sect. 3 presents specific case studies that make use of this framework. Finally, Sect. 4 concludes our work, describing also possible future directions.

2 Adjusting for Opponent Strength

Schedule strength plays a crucial role in the number of games an NFL team wins. During 2015–2016 the easiest schedule was 4 points per game easier than the toughest schedule. This difference of 64 points is worth approximately 2 wins per season. Therefore, it is important to adjust the raw points per play that we showcased in the previous section, based on the strength of the schedule a team faces. For example, let us consider that the average running play gained 0.06 points (and the defense gave up 0.06 points). Suppose Team A averages 0.10 points per run but in each game played against running defenses that gave up 0 points per run. Then clearly Team A's running offense was better than 0.1 points per run, since it faced better than average defense.

We illustrate our adjustment procedure by demonstrating how to compute the adjusted points on running plays - for the rest of the play types (passes, punts, etc.) the method is the same. This method involves first calculating offensive and defensive ratings for the rushing offense and defense of each team. These ratings essentially capture how much better or worse per play a rushing offense/defense is compared to an average rushing offense/defense. We consider the unadjusted points v_i on each running play i to be the score of a single-play game between the offensive team's running offense and the defending team's running defense. Then we can predict for play i the point margin gained for the offense as:

$$\hat{p}_i = m_{rush} + r_{o_i} + r_{d_i} + \epsilon \tag{1}$$

where:

- $\hat{p}_i$ is the point margin gained for the offense for rushing play i
- m_{rush} is a constant representing the league average points per rush
- r_{o_i} (r_{d_i}) is the offensive (defensive) ability of the offensive (defensive) team of play i
- ϵ is the error term of the prediction

Note here that a positive value for r_{d_i} indicates a below average rushing defense while a positive r_{o_i} indicates an above average rushing offense. To find the rushing (offensive and defensive) ratings for each team we need to solve the following constraint optimization problem:

---

[1] While similar metrics exist, their method is not public [2], while from the limited description they seem to still suffer from schedule strength biases (see Appendix B).

$$\underset{m_{rush}, \mathbf{r}_o, \mathbf{r}_d}{\text{minimize}} \quad \sum_{i=1}^{N} (v_i - (m_{rush} + r_{o_i} + r_{d_i}))^2$$

$$\text{subject to} \quad \sum_{j=1}^{32} n_{o_j} \cdot r_{o_j} = 0, \tag{2}$$

$$\sum_{j=1}^{32} n_{d_j} \cdot r_{d_j} = 0.$$

where n_{o_j} and n_{d_j} are the number of offensive and defensive rushing plays for team j, v_i is the raw point value for play i and N is the total number of rushing plays. The two constraints normalize the weighted (by the number of plays) average of the teams rushing offensive and defensive abilities to both be equal to 0.

After solving optimization problem 2 we can use the team's rushing ratings to obtain the adjusted point value for each rushing play. The adjustment process will provide two adjusted point values, one from the standpoint of the offense and one from the standpoint of the defense. As it will become evident from our description below this is rational. Let us assume that a rushing play of team A against team B provided an unadjusted value of v. If the rushing defense ability of team B is r_{d_B}, then the adjusted value for the offense is $v - r_{d_B}$, i.e., we simply adjust the value of the play by subtracting the defensive team's rushing ability. Similarly, if the team A's offense rushing ability is r_{o_A}, the adjusted value for B's defense is $v - r_{o_A}$. For example, if the Cowboys are playing the Patriots and Ezekiel Elliott has a rushing play with an unadjusted value of 0.40 points and the Patriots' defensive rush ability is -0.05 points per play (which is better than average). Then the adjusted value of the play for the Cowboy's offense would be 0.40 (0.05) = 0.45 points. Conversely, if the Cowboys' offensive rushing ability is $+0.10$ points per play then from the standpoint of evaluating the Patriots defense this play should have an adjusted value of $0.40 - 0.10 = 0.3$ points.

2.1   What Is the Impact of the Value Adjustment?

To better understand the importance of our adjustment mechanism, we calculate the absolute difference between the raw/unadjusted values and the adjusted values for each play and for both the offense and the defense for the season 2014–2016. Overall, the average absolute discrepancy is 0.06 points per offensive play (p-value < 0.001) and 0.07 points per defensive play (p-value < 0.001). Offense includes rushing and passing offense as well as special teams plays, i.e., kicking, punting and field goal attempts. Similarly, defense includes rushing and passing defense as well as special teams plays, i.e., kick and punt returns as well as field goal defense.

This discrepancy might appear small in a first glance, however, its compounding effect on evaluating plays and players can be significant. In a typical game, there are approximately 130 plays (both offensive and defensive). A discrepancy of 0.06 then translates to $130 \cdot 16 \cdot 0.06 = 124.8$ points. Given that 33 points correspond approximately to 1 win (Appendix C), this translates to approximately 3.8 wins for every team not being credited correctly!

Figure 1 presents the solutions to the optimization problem 2 that provides us with the ratings for the teams with respect to the rushing and passing game (both offensively and defensively). The top two figures correspond to the 2016 season, where as we can see Broncos had the best passing defense, while the Rams had the worse passing offense, with the Falcons having the best passing offense with Matt Ryan having his MVP season. With respect to rushing, Bills and Cowboys were the two top rushing teams (LeSean McCoy's and Ezekiel Elliott's performances during that season had a lot to do with this).

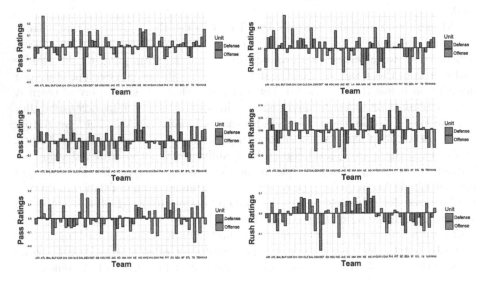

Fig. 1. The rushing and passing ability (both offensive and defensive) as provided by solving the optimization problem 2 for 2014 (bottom) to 2016 (top) NFL seasons

Focusing on the results from the 2016 season it is interesting to notice that the Jets are top-3 rushing defense when we adjust the expected values. If one were to use traditional statistics that are typically used in broadcasting, such as yards/game, points/game etc. the Jets would be not make the top-10 of the list. The reality though is that the Jets had a very good rushing defense in 2016. Points against is also fatally flawed as an indicator of defense ability, since a poor offense often results in many points scored against the team as well. Why should a pick-6 count against the defense? Similarly, why a fumble that gives the opposing offense a short field should assign all the blame to the defense in case of an ensuing score? With expected points, the offense will be held accountable for these situations with negative value plays. Also yards given up don't reflect defenses that "bend but don't break". We hope that in the future more evaluation will happen with improved metrics that utilize the notion of point value per play.

This problem is even more pronounced when trying to evaluate the plays from the special teams. While a kick-off or punt return for a touchdown is undeniably a valuable play, it only captures a specific event (which also is rare). Metrics such as average net yards per punt can be extremely misleading. A 50-yards net punt from your own 45 will pin the opponent deep in their territory, while the same punt from your end zone will give great field position to the opposing offense. However, again in the latter case it is not the special team's fault that the offense couldn't move the ball down the field. So, these subtleties are hard (if not impossible) to be evaluated and quantified with the traditional statistics used by mainstream media. Figure 2 presents the special teams ratings for the 2014–2016 NFL seasons for kickoffs and punts respectively. The offense here refers to the team kicking/punting and ultimately covering, and a positive value indicates a good play. On the contrary the defense refers to the returning team, and a negative value indicates a better than average play.

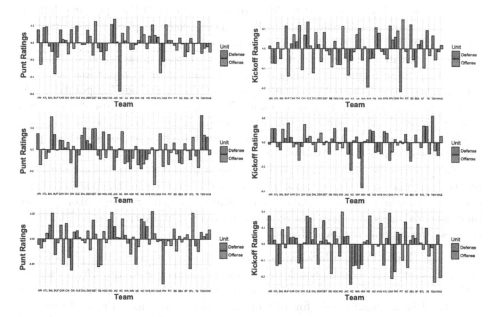

Fig. 2. Adjustment of point values per play allows us to evaluate special teams, an aspect of the game that while analysts agree on its importance it has been ignored in its quantification.

To reiterate, one of the novelties of using expected point values is that we can evaluate the special teams play. Focusing again in the 2016 season we can see that the Cardinals were worse than average in almost all special teams facets of the game (they were slightly above average in kickoff returns). After accounting for the number of plays, we found that Arizona's terrible teams performed 40 points below average and cost Arizona nearly 1.2 wins in 2016! On the contrary, Chiefs special teams had the best punt return unit in the league. Just the punt return game for the Chiefs performed 40 points above average!

3   Case Studies

Passing-vs-Rushing: As it should be evident from the results presented earlier, on average passing plays are much better than running plays: in 2016 running plays averaged 0.069 points per play and pass attempts averaged 0.26 points per play. In prior years, the average points per run and pass attempt are very similar. So why not pass all the time? Isn't this what the analysis suggests? No, and this is where context needs to be integrated in the analytics process to interpret the results. First, the data that were used in the analysis are observational and were obtained based on the teams' specific game plan. If a team calls the same play every time we should expect its efficiency to drop [6] and thus, relying heavily on one type of play might lead to different observations. If the opposing defense is sure the offense will pass, they will be more likely to play a pass oriented defense that reduces passing effectiveness. In fact, we analyzed the expected points gained in 3rd downs from passing and rushing plays when the offense uses the shotgun formation, which is typically indicative of a passing play. For 3rd and less than 5 yards to go, rushing plays gain on average 0.2 points, while passing plays gain on average 0.16 points. This is probably because the shotgun formation signals a probable pass and the defense is *surprised* by a running play (which again might be the reason for the increased efficiency of a rushing play from the shotgun).

Also passing is riskier than running. The standard deviation for the points per rushing play was 0.89 in 2016, while the same number for the passing places was 1.48. Riskier plays may make a successful sustained drive less likely, so the increased riskiness of passes can reduce their overall attractiveness and increase the need for a good rushing offense.

Rushing Gaps: Point values can also be used to evaluate rushing plays based on the rushing hole. During the 2014–2016 seasons, approximately 25% of all the rushing attempts where were up the middle, while each of the rest rushing gaps accounted for approximately for 12% of the rushing attempts. Figure 3 presents the average adjusted points per rush for the offense based on the location of the rushing gap. As we can see rush attempts towards the outside provide more value as compared to rushes up the middle! Of course, this can either be by design or because things up the middle break down and the running back is able to find space towards the outside.

One can also solve optimization problem 2 to obtain offensive and defensive ratings for each team and for each type of run. This can give teams several insights with regards to either their personnel and/or their opponent's personnel. For instance, we have obtained the (offensive and defensive) rushing ratings per

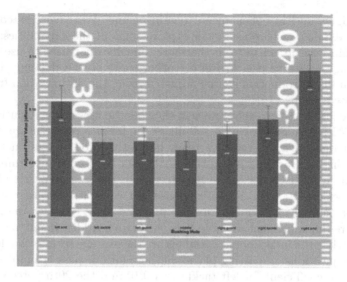

Fig. 3. Rushing attempts closer to the middle provide less expected points per run as compared to attempts towards the end rushing gaps.

gap for the 2016 season (results are omitted due to space limitations), which show that the Cardinals were the best team in defensing runs through the right guard, but they were not as successful through the left guard.

4 Discussion and Conclusions

In this study, we have introduced a framework for adjusting point for opponent strength the expected point values per play in NFL. Assigning a point value for a play is not a new concept and in fact there are several approaches that have been proposed for identifying these values, which we discussed earlier. However, none of them adjusts the values for opponent strength. Allowing 0.5 points on a rushing play from LeVeon Bell is certainly not as *bad* as allowing 0.5 points on a rushing play from Paul Perkins. The main contribution of our work is the design of the adjustment process that accounts for opposition strength. In particular, we formulate a constrained optimization problem, whose solution provides us with team ratings for each unit (i.e., rush offense, rush defense, punt return etc.) that can then be used to adjust the raw point values for each play. We further show, how these values can used to evaluate special teams' units. This has been identified as a crucial aspect for the success of a team but there has been little if any effort in evaluating and quantifying special teams' performance. One can easily utilize our results to develop a customized database which allows coaches to determine the effectiveness of their team's (and their opponents) various offensive and defensive plays in any game situation. For example, the 2016 Packers were a poor running team outside of the red zone. However, inside the red zone their rushing game was great, probably due to defenses being worried about the great

Aaron Rodgers and his phenomenal passing game. We have created such an interactive database where users can explore some of the capabilities: https://athlytics.shinyapps.io/nfl-pv/. Furthermore, one does not have to use the WCS for obtaining the raw values. Any of the other (raw) point value models that we discussed in the introduction can be adjusted using the optimization framework we introduced in this work.

We have just scratched the surface in showing the uses of the Adjusted Points per Play (APP). A number of possibilities for further research are open. For example, if the coaching staff of a team has ratings for players on their ability on running and passing plays, then APP can be used to determine a football version of WAR (Wins Above Replacement). Let's assume that players are rated on a scale 1–100. For a team's offense, we would run two regression models (one for passing and one for rushing plays). Let us consider the regression model where the dependent variable Y is the team's Added Points per 100 Pass Attempts, while the independent variables are the average team rating for each of 11 offensive positions on passing plays. Suppose for our pass attempt regression the regression coefficient for left tackle was 0.10 and the 20th percentile of left tackle's passing rating is 30 (we consider the 20th percentile to be the replacement level [3,9]). Then a left tackle with a passing rating of 90 who played on 500 pass attempt snaps would have added $(90 - 30) \cdot 5 \cdot (0.10) - 30$ points above replacement on passing plays (note that the dependent variable is defined per 100 pass attempts). By a similar logic, using the rushing regression, assume the left tackle added 5 points on rushing plays. Then this left tackle added 35 points above replacement, or approximately 1 WAR. This type of analysis would allow NFL teams to move closer to identifying the real value of a player. Of course, building these models requires accurate player ratings which teams might have (either through third parties such as Pro Football Focus or through their own efforts) but unfortunately it is not available to us.

Finally, the adjustment framework we introduced in this work can be useful in evaluating college prospects. This is rather important since college football exhibits even higher degree of uneven strength schedule. In particular, if we use college football play-by play data and determine an Adjusted Points Per Play Value for each NCAA team's rushing and passing defense, then we could adjust a college player's contributions based on the strength of the defenses they faced. This can further enhance the efforts of NFL teams to better evaluate draft picks based on their college performance.

Appendix A   Intuition Behind the WCS Expected Points Model

In this appendix, we are going to present the main idea behind the WCS model for the value of each play. Let us consider for simplicity a 7-yard field as shown in the figure below. The rules of this 7-yard field football are simple and similar to the original game. We have one play to make a first down. It takes only 1 yard to get a first down. We have a 50% chance of gaining 1 yard and 50% chance of gaining 0 yards on any play. When we score, we get 7 points and the other team gets the ball 1 yard away from their goal line.

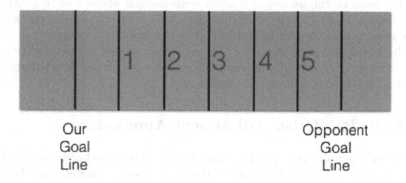

Fig. 4. A 7-yard football field for illustrating the WCS point value model.

If we assume $V(i)$ is the expected point margin by which we should win an infinite game if we have the ball on yard line i, then we can use the following systems of equations to solve for every $V(i)$:

$$V(1) = 0.5 \cdot V(2) - 0.5 \cdot V(5) \tag{3}$$

$$V(2) = 0.5 \cdot V(3) - 0.5 \cdot V(4) \tag{4}$$

$$V(3) = 0.5 \cdot V(4) - 0.5 \cdot V(3) \tag{5}$$

$$V(4) = 0.5 \cdot V(5) - 0.5 \cdot V(2) \tag{6}$$

$$V(5) = 0.5 \cdot (7 - V(1)) - 0.5 \cdot V(1) \tag{7}$$

To derive the above equations we condition on whether we gain a yard or not. For example, for Eq. (3), suppose we have the ball on the 1-yard line. Then with probability 0.5 we gain a yard (and the situation is now worth V(2)), while with probability 0.5 we do not gain this yard and the other team gets the ball one yard away from our goal line. Hence, their worth is V(5), which for our team means that this situation is −V(5). Equations (4)–(7) are obtained in a similar manner. For Eq. (7) we have to note that if we gain a yard (which happens with probability 0.5), we gain 7 points and the opponent will get the ball on its own 1-yard line (a situation worth V(1) to the opponent). Therefore, with probability 0.5 we will gain 7-V(1) points. By solving these equations we get: V(1) = −5.25, V(2) = −1.75, V(3) = 1.75, V(4) = 5.25 and V(5) = 8.75. Thus, in this simplified football game each yard line closer to out goal line is worth 3.5 points (or half a touchdown) (Fig. 4).

Adapting this method to the actual football game requires identifying the "transition probabilities" that indicate the chance of going from say, first and 10 on your own 25-yard line to second and 4 on your own 31-yard line. These transition probabilities are difficult to estimate simulations through the *Pro Quarterback* game[2] that allows the offensive team to choose one of 12 plays and the defensive team to choose one of 9 plays were used; it allowed us to generate a distribution of yards gained, which can then be used to obtain the corresponding recursive equations (similar to (3)–(7)). Today one could replicate this model using data from play-by-play logs, or (to avoid the sparsity problem of the situations) video game simulations in much of a similar manner that Pro-Quarterback was used by Winston, Cabot and Sagarin.

Appendix B   Other Adjustment Approaches

While there is not any academic literature (to the best of our knowledge) tackling the adjustment for schedule strength, there are sports websites (e.g., Football Outsiders) that have their own defense-adjusted metrics. The exact description and methodology behind the calculations of these metrics is not public (possibly available through subscription), and only a short description is available: *"passing plays are also adjusted based on how the defense performs against passes to running backs, tight ends, or wide receivers. Defenses are adjusted based on the average success of the offenses they are facing."* [2]. However, it is not clear how this average success (both offensive and defensive) is calculated. If it is a simple average, then this is still biased from the competition faced by the team. Simply put, one can imagine that our optimization approach resembles a very large number of iterations where the team ratings are re-calculated, while using a simple average is one single iteration of a similar process.

Appendix C   33 Points Equal 1 Win in NFL

Where did the equivalence of 33 points to 1 win came from? An easy way to understand this is by examining the relationship between the point differential for a team (i.e., points for minus points against) for a whole season with the total number of wins for the team. Using the data for the season 2014–2016 we obtain the following relationship. Fitting a linear regression model on the data we obtain an intercept of 7.9 and a slope of 0.03. Simply put, a team with 0-point differential is expected to have an 8-8 record, while every increase in the point differential by 1 point, corresponds to 0.03. Equivalently, 1 win corresponds to $\frac{1}{0.03} \approx 33$ points (Fig. 5).

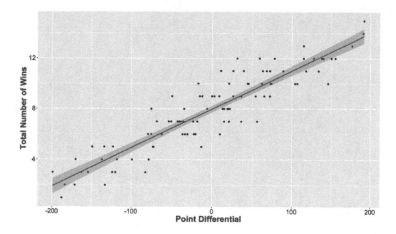

Fig. 5. There is a linear relationship between the point differential of a team and its total number of wins.

References

1. Expected points (ep) and expected points added (epa) explained. http://archive. advancedfootballanalytics.com/2010/01/expected-points-ep-and-expected-points. html. Accessed 19 June 2018
2. Football outsiders: Methods to our madness. https://www.footballoutsiders.com/ info/methods. Accessed 19 June 2018
3. Prospectus toolbox: Value over replacement player. https://www. baseballprospectus.com/news/article/6231/prospectus-toolbox-value-over- replacement-player/. Accessed 19 June2018
4. Carroll, B., Palmer, P., Thorn, J.: The Hidden Game of Football. Warner Books, New York (1989)
5. Carter, V., Machol, R.E.: Operations research on football. Oper. Res. **19**(2), 541– 544 (1971)
6. Pelechrinis, K.: The passing skill curve in the NFL. In: The Cascadia Symposium on Statistics in Sports (2018)
7. Shapley, L.S.: Stochastic games. Proc. Natl. Acad. Sci. **39**(10), 1095–1100 (1953)
8. Winston, W., Cabot, A., Sagarin, J.: Football as an infinite horizon zero-sum stochastic game. Technical report (1983)
9. Winston, W.L.: Mathletics: How Gamblers, Managers, and Sports Enthusiasts Use Mathematics in Baseball, Basketball, and Football. Princeton University Press, Princeton (2012)
10. Yurko, R., Ventura, S., Horowitz, M.: nflwar: A reproducible method for offensive player evaluation in football. arXiv:1802.00998 (2018)

Individual Sports

Real-Time Power Performance Prediction in Tour de France

Yasuyuki Kataoka[1]([✉]) and Peter Gray[2]

[1] NTT Innovation Institute Inc., East Palo Alto, CA 94089, USA
`kataoka.yasuyuki@ntti3.com`
[2] Dimension Data Australia, Port Melbourne, VIC 3207, Australia
`peter.gray@dimensiondata.com`

Abstract. This paper introduces the real-time machine learning system to predict power performance of professional riders at *Tour de France*. In cycling races, it is crucial not only for athletes to understand their power output but for cycling fans to enjoy the power usage strategy too. However, it is difficult to obtain the power information from each rider due to its competitive sensitivity. This paper discusses a machine learning module that predicts power using the GPS data with the focus on feature design and latency issue. First, the proposed feature design method leverages both hand-crafted feature engineering using physics knowledge and automatic feature generation using autoencoder. Second, the various machine learning models are compared and analyzed with the latency constraints. As a result, our proposed method reduced prediction error by 56.79% compared to the conventional physics model and satisfied the latency requirement. Our module was used during the *Tour de France 2017* to indicate *an effort index* that was shared with fans via media.

Keywords: Machine learning · Recurrent neural network · Autoencoder · Real-time system · Spatiotemporal data · Sports analytics

1 Introduction

Measurement of muscle power in cycling has become an attractive tool for professional riders, coaches, and amateurs to improve the riding performance. For instance, it allows coaches to help monitor the training effectiveness when combined with heart rate measurement. It also allows riders to tactically determine energy use by analyzing other's muscle fatigue level. Moreover, it helps audiences to enjoy the competition by monitoring the rider's performance.

However, there are two issues in power data collection in cycling sports. First, power sensors tend to be expensive. Reducing cost of power meters means that they are becoming more accessible to competitive and even recreational amateur cyclists. Second, data on muscle usage is usually highly confidential, and it is not easily accessible. Although most professional teams attach power meters, the performance information is usually confidential within the team.

© Springer Nature Switzerland AG 2019
U. Brefeld et al. (Eds.): MLSA 2018, LNAI 11330, pp. 121–130, 2019.
https://doi.org/10.1007/978-3-030-17274-9_10

The purpose of this paper is to discuss the real-time machine learning system that predicts muscle power in cycling competition to enable people to get access to power performance information in *Tour de France*[1].

Conventionally, the power data is analyzed by the physics model which heavily relies on not data-driven but model-driven approach. This approach heavily depends on the physical constants, which tends to be less accurate. The challenge of the data-driven approach is collecting the labeled data with power information along with GPS data. Fortunately, in cooperation with one of the professional cycling teams and *Dimension Data's data analytics platform*[2] which collects GPS data from all riders, we obtained the labeled dataset for this purpose.

This paper proposes the data-driven power prediction that fuses the physics model with 1. feature design method and 2. real-time machine learning model analysis. First, the proposed feature design method leverages both hand-crafted feature engineering using physics knowledge and automatic feature generation using deep autoencoder. Beyond the previous studies of muscle fatigue analytics for cyclists, the feature inspired by deep learning enables trajectory patterns to be embedded into the model. This generated feature allows us to implicitly consider the rider's behavior such that the power use is loosened in the context of turning a sharp corner on a downhill slope. Second, the tree-based machine learning models and time-series deep learning models are compared regarding latency and error rate.

As a result, our ultimate model reduced prediction error by 56.79% compared to the conventional model-based model that depends on the prior knowledge of physics. Our Machine Learning module was used during the *Tour de France 2017* in a real-time manner to create *an effort index*, the power indicator, that was shared with fans via social media. Our proposed method can be used for amateur riders too who want to know the power performance but does not want to purchase a real power meter which tends to be very expensive.

2   Related Work

The performance of cycling riders has been studied across various academic fields [1]. Among them, this paper focuses on muscle fatigue and addresses the problem of predicting it by machine learning.

Fatigue Analytics. In the study of muscle fatigue in cycling, models considering various factors have been proposed. For example, one proposed model considers physiological, biomechanical, environmental, mechanical and psychological factors and integrates them into nonlinear complex system models [2]. While many researchers take the model-based approach [3,4], this paper focuses on a data-driven approach using the limited available dataset such as GPS.

---

[1] *Tour de France* is one of the three major European professional cycling stage races in road bicycle racing. https://en.wikipedia.org/wiki/Tour_de_France.

[2] *Dimension Data's data analytics platform*, https://www2.dimensiondata.com/tourdefrance/analytics-in-action.

The fatigue analytics is, in general, used for performance or safety improvement in the sports industry. One example is finding the relaxation place during a race [5]. This paper aims for fan engagement, which is also important in sports industry from the business perspective.

Machine Learning Application Using Cycling Data. The most famous machine learning applications using cycling GPS data is the transportation mode prediction: classifying user's activity to bicycle, car, train, walk, run and others [6]. One study reported that the generated feature from GPS trajectory using Deep Learning improves accuracy [7]. This is because the Deep Learning automatically captures important features which are difficult to be designed explicitly by hand-crafted feature engineering [8]. In our limited but best knowledge, there is no research report about muscle fatigue prediction using both hand-crafted feature and generated feature by deep learning in cycling sports.

3 Dataset

Input Data: GPS and Wind Sensor. *Dimension Data's data analytics platform* has a live GPS tracking system. This system provides the GPS tracking of position and speed for all riders at a 1 Hz frequency from the GPS sensors mounted under the bicycle saddle. This data is processed in real time, and enriched to calculate key metrics such as distance to finish, position in the race, time gaps, clustering of individual riders into groups, and the additional environmental data such as the current gradient of the road and the wind conditions.

Labeled Data: Power Sensor. Power is the measurement of how much force is being pushed through the pedals by the rider and is measured using dedicated sensors usually built into cranks, pedals or rear wheel hub. Most power meters connect wirelessly to the rider's bike computer allowing them to monitor their power output during a training session or race and manage their effort accordingly. In this project, a training dataset was obtained from one of the professional cycling teams in previous professional races. This dataset includes the data *Dimension Data's data analytics platform* provides as well as the power sensor data in accordance with the time stamp. The total count of the valid labeled data is 68849 after cleansing the dataset.

4 Methodology

This section mainly describes how the machine learning model is designed for power prediction with the focus on 1. feature design method and 2. real-time machine learning model analysis. In the feature engineering part, the hand-crafted feature is designed by mechanical factors using fundamental physics. Also, the generated feature by autoencoder is concatenated to the feature space. In the regression model, the various machine learning models are introduced with the arguments of advantages and latency perspectives.

4.1   Feature Engineering

Our proposed feature design is shown in Fig. 1: a hand-crafted feature inspired by physics knowledge and an automatically generated feature using deep learning.

Hand-Crafted Feature. Rider's power is physically determined by four factors:

1. *friction with the ground* denoted as $P_f = C_f v_b mg$, where C_f is the friction coefficient, v_b is bicycle velocity, m is total mass, and g is standard gravity.
2. *wind resistance* denoted as $P_w = 1/2 C_d A \rho (v_b - v_w)^2$, where C_d is drag coefficient, A is frontal area, ρ is air density, and v_w is wind velocity.
3. *kinetic energy* denoted as $P_k = \frac{m}{2 \Delta T}(v_n^2 - v_p^2)$, where v_n is the velocity at $t = now$, and v_p is the previous velocity at $t = now - \Delta T$.
4. *potential energy* denoted as $P_p = mg \frac{\Delta h}{\Delta T}$, where ΔT is sampling time interval, and h is height variation within ΔT.

Each coefficient are surveyed in various reports[3]. However, we realized that the power calculated by these values is greatly different from the data from real sensors for our dataset. Therefore, the power prediction model is designed by machine learning, which identifies the desired coefficients to fit with our professional rider's dataset.

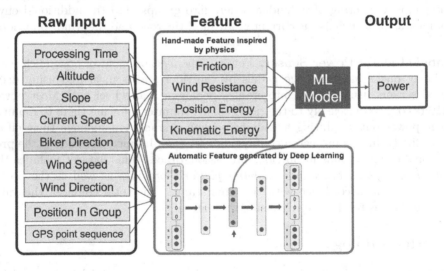

Fig. 1. Machine Learning pipeline: both hand-crafted feature by physics and automatic feature generated by Deep Learning is concatenated for Machine Learning model.

---

[3] One example of the physical constants https://www.cyclingpowerlab.com/Cycling Aerodynamics.aspx.

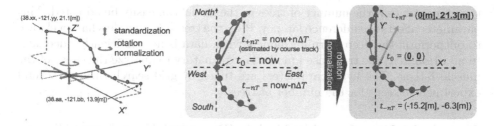

Fig. 2. Data normalization for GPS trajectory: The main idea of this preprocess is to align the direction for all of the data.

Generated Feature by Deep Learning. It is assumed that the rider's power use is influenced by past and future trajectory of a rider. For example, it is observed that the pedal is stopped in the context of turning a sharp corner on a downhill slope. When fusing GPS trajectory data to the model, we found two issues: 1. direction diversity, 2. high dimensionality.

First, normalization of the trajectory direction is performed in the (x, y) plane. As shown in Fig. 2, the rotation transformation is applied so that the direction of the vector towards the position after $N\Delta T$ seconds corresponds to the positive direction of the y-axis. Here, the future GPS points are predicted based on the assumption that the current speed is maintained along with the course track. After applying rotation normalization, standardization is applied for each x, y, and z-axis.

Second, a dimensional reduction is applied by various autoencoders such as denoising autoencoder [9], deep autoencoder [10], and stacked deep autoencoder [11]. The input vector is the GPS points from $t - N\Delta T$ to $t + N\Delta T$, where each GPS point has x, y, and z value. This ends up a total $3(2T+1)$ dimension for input space, which tends to be sparse. These deep layers make it accurate to restore the input, meaning that implicit but powerful feature of the trajectory is extracted automatically. In this paper, the compressed feature vector by deep autoencoder is called the 'embedded trajectory feature'. This embedded trajectory feature is concatenated to the hand-crafted feature.

4.2 Regression Model

The challenge of the model choice is, in general, to optimize the model with respect to latency and error rate. The latency issue is critical in this real-time power prediction application, because it must predict power for each of 198 riders within one second. In the case of simple scenario by one machine, it is necessary to complete one prediction approximately in 5 ms. Within this 5 ms, the following process needs to be completed: subscribe incoming data, compute feature, run inference, and send outcome to the database. Although the distributed computing can solve this challenge, we set the latency requirement to 2 ms in model optimization task.

Tree-Based Models. Random Forest [12] and XGBoost [13] are considered as part of the regression model candidates. The advantage of the decision tree type model is that the number of trees in the model can easily be adjusted. This parameter affects the inference latency. Plus, the tree-based models have a chance to outperform the deep learning models when data is not sufficiently adequate. In addition to it, the tree-based model is explanatory to analyze the cause of the muscle fatigue. The hyperparameters are tuned by grid search through several experiments except for the number of trees.

Time-Series Deep Learning Models (Recurrent Neural Net). Stacked Long Short-Term Memory (LSTM) [14] and Gated Recurrent Units (GRU) [15] are considered as part of the regression model candidates from Recurrent Neural Net (RNN) models. The advantage of RNN is that predictive performance may outperform other models by extracting effective features over time-series information. After several experiments, some hyperparameters are fixed, e.g., the number of the past time-series data $= 10$, dropout ratio $= 0.4$. In this paper, the number of the layer numbers is treated as hyperparameter.

5   Result

First, this section quantitatively evaluates the accuracy of the power prediction regarding feature engineering, embedded trajectory feature, and regression models. Moreover, this section qualitatively evaluates the impact of the use of this machine learning model on fan engagement at the *Tour de France 2017*.

In the evaluation, *stratified* 5-fold cross validation is applied, because the dataset is imbalanced data. The metrics for the evaluation is mean absolute error (MAE), which computes the absolute value between the predicted value and the ground truth.

5.1   Evaluation on Feature Engineering

The purpose of this section is to analyze the effect of feature engineering by both hand-crafted features inspired by physics and generated feature by trajectory embedding autoencoder. In this comparison analysis, the following four different model types are considered:

1. **M1**: Physic-based Model (baseline)
   M1 is the conventional power model, the sum of four power factors $P = P_f + P_w + P_k + P_p$, where coefficients are determined by other articles.
2. **M2**: Data-Driven Model without feature engineering
   M2 is a machine learning model without any additional feature engineering. This simply uses raw input described in Fig. 1.
3. **M3**: M2 + hand-crafted Feature
   In addition to M2, M3 considers the hand-crafted feature designed in Sect. 4.1.
4. **M4**: M3 + Embedded Trajectory Feature by denoising stacked autoencoder
   In addition to M3, M4 considers the embedded trajectory feature designed in Sect. 4.1. The parameters $N = 5$, dimension of autoencoder's layers $= [33, 20, 10, 20, 33]$.

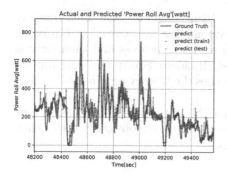

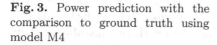

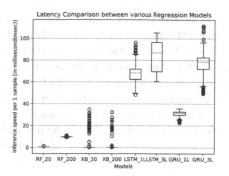

Fig. 3. Power prediction with the comparison to ground truth using model M4

Fig. 4. Latency evaluation (RF = Random Forest, XB = XGBoost, _X= # of estimator or layers)

The result is shown in Table 1. Although the prediction may be difficult in high power range (>400 watt) or low power range (<100 watt) due to the imbalanced training dataset, Fig. 3 indicates our proposed model can work accurately in these challenging ranges too. Then, the comparative evaluation is shown in Table 1. Our proposed method, M4, outperforms the simple model-based model using only physics by 56.79% error reduction in MAE. Compared to M2, the simple data-driven model, our feature design improves machine learning model by 35.40% error reduction in this experiment. Thus, both hand-crafted feature and embedded trajectory feature should help to capture important factors to predict power use in cycling.

5.2 Evaluation on Regression Models

Inference Latency. In this experiment, the inference process was run on 198 samples data on GPU server (Tesla K80) whose status is idle except for this experiment. Note that the computation of 198 samples by matrix must not be run at once, because it needs to be done one by one in the real scenario. The results of the inference latency analysis are shown in Fig. 4 by box plot.

The best latency performance measured by median is XGBoost with the 200 trees. While the average performance of XGBoost outperforms other regression

Table 1. Performance comparison between four different feature engineering by MAE.

Model type	MAE (Train)	MAE (Test)	Error reduction to baseline
M1(Baseline)	-	139.17	-
M2	37.55	89.90	35.40%
M3	24.36	66.82	51.99%
M4	21.86	60.13	**56.79%**

models, the latency widely varies and takes +10 ms for some cases. This anomaly causes the negative impact on the backend system. One negative effect is missing values. The backend system terminates the inference process and then returns NaN for some cases. Contrary to XGBoost, Random Forest fairly performed stably. Random Forest with 20 tree trees satisfies the latency requirement, which is set to be 2 ms as described in Sect. 4.2.

The time-series deep learning models, LSTM and GRU did not satisfy our latency requirement. When the multiple layers are stacked, obviously the latency gets worse due to the additional computation.

Error Rate - MAE. The result is shown in Table 2. Among the tree-based models, XGBoost (n_est = 200) has shown the best performance with the satisfactory latency on average. However, it has the problem with the unstable latency. Thus, the Random Forest (n_est. = 20) is considered to be the best regression model in our practical situation. The time-series deep learning models, LSTM and GRU, turned out not to be better than tree-based models. It surely ends up underfitting. One famous way to avoid underfitting is to change the model to deeper structure. However, this approach did not work in this experiment. It may be because of the less labeled dataset to train the effective deep learning model. Even if it may get outperform tree-based models by adding more dataset available in future, the unstable and longer latency performance would not be solved.

5.3 Qualitative Analysis - Real Deployment in Tour de France 2017

Our proposed machine learning model was successfully deployed in Dimension Data's data analytics platform. In *Tour de France 2017*, the analytics platform team decided to use the output of power prediction as *an effort index* which indicates the power level from 1 to 10 for better visualization to fans and for respects to rider's semi-private data. This was the first trial to compute the effort

Table 2. Performance comparison between regression models by MAE and latency

Model type	MAE (Train)	MAE (Test)	Average latency [ms]
Random Forest (n_est. = 20)	24.35	**66.81**	**1.09**
Random Forest (n_est. = 200)	23.48	**66.02**	9.76
XGBoost (n_est = 20)	32.02	70.72	2.94
XGBoost (n_est = 200)	0.37	**63.97**	1.07
LSTM (1 Layer)	289.18	296.01	67.79
LSTM (3 Layer)	280.49	289.21	84.22
GRU (1 Layer)	289.78	290.78	30.72
GRU (3 Layer)	280.49	289.22	77.02

index in the history of *Tour de France* or any other cycling competition in our best knowledge.

Figure 5 shows one of the real-time visualization tools using the power prediction. This tool enables a user to compare the performance of two different groups at specific time range: e.g., Peloton vs Front Group in the past 5 min. This can visualize how enthusiastically peloton saves energy during a race or tries to catch up the front group.

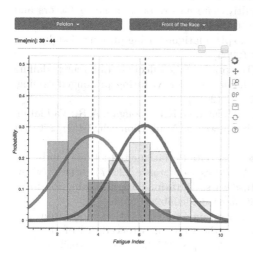

Fig. 5. Real-time visualization of power distribution for two group: x-axis is fatigue index, y-axis is probability

Fig. 6. Example of social media exposure: visualization of the winner's performance in accordance with the terrain variation

Figure 6 shows the social media exposures of our technology: one tweet by Dimension Data that describes how the winner on stage 18 expends energy in accordance with terrain variation of the course. This graph indicates how the winner saved at downhill before the final uphill and used the peak effort at the end of the race.

6 Conclusion

This paper presented a machine learning application of power prediction used in *Tour de France 2017*. The characteristic approach of this paper is the feature design combined with both hand-crafted feature based on physics and generated features based on deep autoencoder using GPS trajectory. As a result, the error (MAE) rate is reduced by 56.79% compared to the physical model, and by 21.39% compared to the basic machine learning model. Moreover, several regression models are investigated regarding error rate and latency. This power prediction application contributes to fan engagement in cycling sports, as evidenced by social media. In the future, we plan to gather amateur riders' datasets for further sensorless power prediction products at a lower price than the power meters, which are often unaffordable for the ordinary consumer.

References

1. Castronovo, A.M., Conforto, S., Schmid, M., Bibbo, D., D'Alessio, T.: How to assess performance in cycling: the multivariate nature of influencing factors and related indicators. Front. Physiol. **4**, 116 (2013)
2. Abbiss, C.R., Laursen, P.B.: Models to explain fatigue during prolonged endurance cycling. Sports Med. **35**(10), 865–898 (2005)
3. Theurel, J., Crepin, M., Foissac, M., Temprado, J.J.: Effects of different pedalling techniques on muscle fatigue and mechanical efficiency during prolonged cycling. Scand. J. Med. Sci. Sports **22**(6), 714–721 (2012)
4. Martin, J.C., Milliken, D.L., Cobb, J.E., McFadden, K.L., Coggan, A.R.: Validation of a mathematical model for road cycling power. J. Appl. Biomech. **14**(3), 276–291 (1998)
5. Kataoka, Y., Junkins, D.: Mining muscle use data for fatigue reduction in IndyCar. In: 11th Annual MIT Sloan Sports Analytics Conference (2017). https://www.semanticscholar.org/paper/Mining-Muscle-Use-Data-for-Fatigue-Reduction-in-Kataoka-Junkins/abac602b26f7dd97655aa6443efde0ab9e90b186
6. Zheng, Y., Liu, L., Wang, L., Xie, X.: Learning transportation mode from raw GPS data for geographic applications on the web. In: Proceedings of the 17th International Conference on World Wide Web, pp. 247–256. ACM, April 2008
7. Endo, Y., Toda, H., Nishida, K., Ikedo, J.: Classifying spatial trajectories using representation learning. Int. J. Data Sci. Anal. **2**(3–4), 107–117 (2016)
8. LeCun, Y., Bengio, Y., Hinton, G.: Deep learning. Nature **521**(7553), 436 (2015)
9. Vincent, P., Larochelle, H., Bengio, Y., Manzagol, P.A.: Extracting and composing robust features with denoising autoencoders. In: Proceedings of the 25th International Conference on Machine Learning, pp. 1096–1103. ACM, July 2008
10. Hinton, G.E., Salakhutdinov, R.R.: Reducing the dimensionality of data with neural networks. Science **313**(5786), 504–507 (2006)
11. Vincent, P., Larochelle, H., Lajoie, I., Bengio, Y., Manzagol, P.A.: Stacked denoising autoencoders: learning useful representations in a deep network with a local denoising criterion. J. Mach. Learn. Res. **11**(Dec), 3371–3408 (2010)
12. Liaw, A., Wiener, M.: Classification and regression by randomForest. R news **2**(3), 18–22 (2002)
13. Chen, T., Guestrin, C.: Xgboost: a scalable tree boosting system. In: Proceedings of the 22nd ACM SIGKDD International Conference on Knowledge Discovery and Data Mining, pp. 785–794. ACM, August 2016
14. Hochreiter, S., Schmidhuber, J.: Long short-term memory. Neural Comput. **9**(8), 1735–1780 (1997)
15. Chung, J., Gulcehre, C., Cho, K., Bengio, Y.: Gated feedback recurrent neural networks. In: International Conference on Machine Learning, pp. 2067–2075, June 2015

Automatic Classification of Strike Techniques Using Limb Trajectory Data

Kasper M. W. Soekarjo[1], Dominic Orth[1,2,3(✉)], Elke Warmerdam[1], and John van der Kamp[1,2,3]

[1] Faculty of Behavioural and Movement Sciences, Vrije Universiteit Amsterdam, Amsterdam, Netherlands
d.o.orth@vu.nl
[2] Amsterdam Movement Sciences, Amsterdam, The Netherlands
[3] Institute of Brain and Behavior, Amsterdam, The Netherlands

Abstract. The classification of trajectory data is required in a wide variety of movement tracking experiments. Automatic classification using machine learning techniques has the potential to greatly increase efficiency and reliability of these studies. Here, we apply supervised classification algorithms on a dataset obtained through a kickboxing experiment to classify the limb and technique that was used for each strike as well as the expertise of the person performing the strike. Beginner and expert kickboxers were asked to strike a boxing bag from several distances, producing a dataset of approximately 4000 strike trajectories. These trajectories were classified using the K-nearest neighbours (KNN) and multi-class linear support vector classification (SVC). We show that both of these algorithms are capable of correctly classifying the limb used for the strike with ~99% prediction accuracy. Both algorithms could classify the techniques used with ~86% accuracy. The accuracy of technique classification was improved even further by applying hierarchical classification, classifying techniques separately for each limb. Only 10% of the dataset was required as training set to approach the observed prediction accuracy. Finally, KNN was capable of classifying the strikes by skill level with 73.3% accuracy. These findings demonstrate the potential of using supervised classification on complex limb trajectory datasets.

Keywords: Machine learning · Strike technique classification · Limb trajectory analysis · Combat sport

1 Introduction

The analysis of limb trajectories has long been an area of interest in order to improve knowledge about performance, injury prevention, and understanding the functionality of movement variability [4,5,8,9,13]. However, analyzing complex

This project received the support of the Nederlandse Organisatie voor Wetenschappelijk Onderzoek (NWO), reference number: 464-15-130.

ⓒ Springer Nature Switzerland AG 2019
U. Brefeld et al. (Eds.): MLSA 2018, LNAI 11330, pp. 131–141, 2019.
https://doi.org/10.1007/978-3-030-17274-9_11

strike patterns often requires manual labeling of the techniques that are used [1]. This increases the complexity and labour required to analyse the data, but was necessary for reliable analysis. Recently, [7] used a supervised classification method for automatically classifying punching techniques from video images. Their work was focused on boxing, so only punches were analysed. Here, we apply similar methods to a more complex dataset of beginning and experienced kick boxers striking a boxing bag. This dataset contains both punches and kicks and a variety of different techniques, complicating the task of accurately classifying the strikes. In this paper, we apply supervised classification to a dataset of 4000 strike trajectories using K-nearest neighbor (KNN) classification and multi-class linear support vector classification (SVC). We show that both models are capable of classifying the limb and technique used for the strike using a small fraction of the dataset. Furthermore, we demonstrate that supervised classification can classify strikes according to the skill level of the person performing the strike. This paper serves as a primer to demonstrate the possibilities of applying automatic classification to complex trajectory data.

2   Methods

2.1   Data Collection

Our primary objective was to examine the effect of skill in a task where participants aimed to maximize both the originality and functionality of striking actions in adapting to changes in external constraints (the distance to a target to strike). In doing so, a group of currently active kickboxers ($n = 21$, age = 25.9) and a group of participants naive to combat sports ($n = 21$, age = 25.3) were recruited and participated. The experimental design required participants strike a free hanging boxing bag. The distance to the midline of the bag was manipulated across 10 positions (see Fig. 1). Each distance corresponded to a percentage of the participants leg length (taken from the trochanter major to the ground while standing). The percentage leg lengths were linearly spaced from 0% to 100% and included: 0%; 11.1%; 22.2%; 44.4%; 55.6%; 66.7%; 77.8%, and; 88.9%; 100%. At each position participants were required to strike the bag 10 times consecutively. There was approximately ten seconds rest enforced between each strike. An additional two minutes rest every ten strikes was also enforced which facilitated adequate rest and afforded equipment checks. The position order for each participant was randomized using sampling without replacement. The specific instructions were: Using your hands and your feet, at each position to vary as much as possible across the ten strikes the method used to strike the bag [originality task constraint]. When striking the bag with the given method, do so as hard and as fast as possible [functionality task constraint]. To ensure that participants adapted to the position (and did not simply move to a self-preferred distance before striking), they were also required to stay within a stand-on-able square. The length of each side of the stand-on-able border was set to the participants shoulder width. We also adjusted the height of the centre of the boxing bag (weight = 45 kg) relative to the height of the sternoid process of each participant. The available surface area that could be struck was also body-scaled to the participant such that they were required to strike either above their knee and below their standing height. Participants were also fitted with safety gear gloves and shin guards for their hands and feet. Instrumentation for obtaining the striking limb trajectory included an optical motion capture system (Optotrak 3020, Northern Digital Inc., Canada) which has an accuracy of 0.5 mm. Two cameras were positioned 4 m at either side of the participant perpendicular to the striking direction. Two active infrared markers were secured to each glove and each ankle (directly beneath the malleolus). Two markers were used to ensure visibility should the participant rotate their limb during a strike. The position of each marker was recorded as 3D data points at a sampling rate of 240 Hz. Before the experiment, the coordinate system was defined using a calibration cube with the origin set directly beneath the longitudinal axis of the bag and subsequent marker paths projected in global coordinates (Fig. 1) [14]. In order to obtain a precise moment of contact with the bag a tri-axial accelerometer was attached to the posterior surface of the bag. The accelerometer measured at 1000 Hz and was linked to the optical motion capture system for time synchronisation. Participants were also equipped with inertial measurement units (IMUs) (MTx, Xsens Technologies, Netherlands) at each body segment and sampled at 120 Hz. IMU data was automatically time synchronized with optical data using the peak correlation of a large amplitude movement of the right foot on the vertical displacement (Figure 1A shows the synchronization and projection results).

2.2   Trajectory Extraction and Labeling

The average of each pair of limb position data was taken. A cubic spline was then used to fill missing values and a low pass Butterworth filter applied. The norm of the accelerometer signals was used to automatically detect moment of contact. We then automatically obtained the limb used to strike the bag, taking the limb associated with the highest acceleration norm (taken directly from the IMUs) around the moment of impact. The beginning of the strike was then determined automatically as when the limb used began to travel with a velocity greater than 0.2 m/s toward the bag. All steps were visually verified. The subsequent dataset consisted of 4235 strike trajectories (Fig. 1B–C). Out of this set, techniques were labeled by observing the full body kinematics of each strike (Fig. 1A). For punching techniques labels included: Hook; Straight; Uppercut; Overhand; Superman Punch; Back Fist; Spinning Back Fist. For kicking techniques labels included: Roundhouse Kick; Front Kick; Stamp Kick; Side Horse Kick; Rear Horse Kick; Spinning Kick. During this step the accuracy of the optical motion capture data for each strike was also verified and acceptable trajectories were retained (Fig. 1B-C).

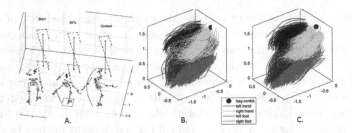

Fig. 1. Retained limb trajectories. (A) Shows the initiation and moment of contact with the target of the projected optical motion capture and inertial measurement unit data obtained for each strike. Beginner (B) and expert (C) trajectories plotted in the global coordinate system (axis values are in metre units).

2.3   Data Cleaning and Analysis

In order to be able to perform and verify supervised classification, each trajectory is required to have outcome labels. Therefore, strikes with missing labels were removed. Furthermore, outcome classes are required to contain at least two members. Therefore, techniques which occurred only once were removed from the dataset. This resulted in a dataset of 3962 strike trajectories usable for classification, out of 4235 originally measured trajectories. The label distribution can be found in Table 1.

Table 1. Technique label distribution in cleaned dataset.

Technique	Count (% of total)
Hook	1385 (35.0)
Straight	908 (22.9)
Roundhouse kick	892 (22.5)
Front kick	354 (8.9)
Uppercut	301 (7.6)
Side horse kick	63 (1.6)
Overhand	28 (0.7)
Spinning kick	17 (0.04)
Stamp kick	6 (0.002)
Rear horse kick	6 (0.002)
Superman punch	2 (0.0005)
Total	3962

There is a strong correlation between subsequent points in the trajectory data. In order to reduce the number of dimensions and remove the correlations between features, principal component analysis was performed on the trajectory data. Each x-, y- and z-coordinate for each of the 100 normalised timesteps were included as features, resulting in 300 features for each trajectory. PCA was performed and the first fifteen principal components were retained, cumulatively explaining 99.99% of variance in the trajectories. This ensured that as much information as possible was retained from the trajectories, while still strongly reducing the amount number of features and removing correlations between features. These fifteen principal components were used as features for all classifications.

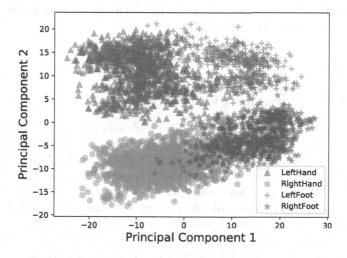

Fig. 2. Principal components of trajectory data. Color indicates which limb was used to perform the strike. (Color figure online)

2.4   Classification Models

Plotting the first two principal components using limb data shows relatively clear separation of limbs used (Fig. 2). This indicates that classification of limbs used should be possible using linear classification. Therefore, two supervised classification algorithms that are capable of detecting linearly separable classes were tested on the dataset. The KNN algorithm is one of the oldest and simplest classification methods available [2]. It classifies data points according to the class of the K nearest neighbors of that datapoint, using the Euclidean distance between points. This allows for the classification of both linearly and non-linearly separable data. Lower values of K generally perform better on stronger separated data and can lead to overfitting on datasets that are not clearly separated, while higher values of K are more robust against outliers and noise but can lead to values on the edges of clusters to be wrongly classified. Higher values of K also require more computational power as more neighbors need to be examined for each datapoint. The optimal value of K is often not known beforehand and has been an ongoing challenge [3,11]. Multi-class linear SVC attempts to classify the data by finding linear vectors that separate the different classes [12]. When performing multi-class classification, two approaches are possible. The simplest method is known as one-against-all, in which one model is trained for each class that separates this class from the rest, resulting in n models for n classes. The other approach is one-against-one, where a model is trained for each possible combination of classes. This results in n x(n−1)/2 models for n classes. The combination of these models is then used to classify each data point. Due to these additional models, one-against-one is more computationally expensive but also less sensitive to unbalanced datasets [6]. All analyses were performed using the scikit-learn package in Python [10].

3   Results

3.1   Limb and Technique Classification

Strike trajectories were classified for which limb was used as well as which technique was performed. Twelve classes of techniques were defined beforehand, of which six were kicks and six were punches. The dataset was split into a training set and a test set which consisted of respectively 75% and 25% of the data. Strikes by the same participant were placed either all in the train set or all in the test set to prevent leakage of information and participant distribution over the train and test sets was randomised for each iteration. In order to determine the optimal number of neighbors K that would result in the highest classification accuracy, K was varied from 1 to 30 and KNN was performed using each value for K. The classification of which limb was used was most accurate for K = 18 (Fig. 3A), while techniques were most accurately qualified with K = 14 (Fig. 3B).

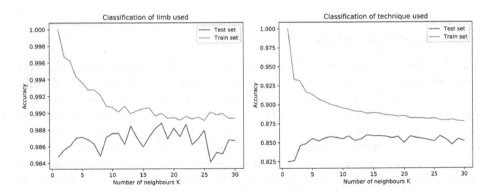

Fig. 3. Classification accuracy of KNN for varying values of K. (A) Shows the classification of which limb was used, (B) Shows classification of strike technique. Orange indicates the accuracy on the training set, blue indicates accuracy on the test set. Values are based on 100 iterations for each value of K. (Color figure online)

Subsequently, linear SVC was performed on the same dataset. Comparison of the classification accuracies of the models showed nearly identical results, with SVC scoring marginally higher for both limb and technique classification (Table 2). The high accuracy of linear SVC indicates that both the limb and the technique classes are linearly separable. This is in accordance with the fact that different limbs and techniques are spatially separated in the trajectory data.

Table 2. Classification accuracy of different models.

	Limb accuracy	Technique accuracy
KNN	98.9%	86.0%
SVC	99.4%	86.7%

3.2   Minimal Required Training Set

After determining that the trajectory data can accurately be classified into the correct limb and technique classes, we attempted to determine the smallest training set size that can still provide accurate classifications. We varied the size of the training set from 1% to 75% of the dataset and assigned the remainder of the data to the test set (Fig. 4). SVC again slightly outperformed KNN for all measurements. For both models, accuracy improvement started leveling off after assigning approximately 10% of the dataset (∼400 strikes) as training data, indicating that this fraction was sufficient to approach the classification accuracy that was originally obtained using 75% of the data for training.

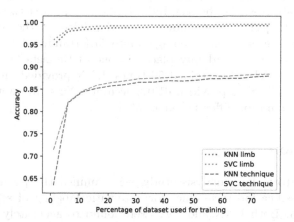

Fig. 4. Accuracy of classification for various training set percentages. Dotted lines show accuracy of limb classification, dashed lines show technique classification. Values are based on 50 iterations for each training set percentage.

3.3 Hierarchical Classification

[7] found that technique classification could be improved by using a hierarchical model, classifying strikes first on limb level and subsequently on technique level. Applying a similar technique, we separated the strike trajectories according to limb classification and then classified the trajectories into technique classes for each limb separately using linear SVC. Accuracy of technique prediction increased for all four limbs (Table 3). These findings confirm the merits of using hierarchical classification for technique recognition. It should be noted that all participants in the experiment were right-handed.

Table 3. Technique label in cleaned dataset.

Technique	Count (% of total)
Right hand	89.8
Left hand	89.1
Right foot	91.1
Left foot	88.0

3.4 Classification of Skill Level

Finally, we explored the capability of the models to predict the skill level of the person performing the strike. Separation of skill level based on trajectory data is less likely to be based on spatial separation, as both beginners and experts used all four limbs to strike the bag. Indeed, coloring the first two principal components of the trajectory data shows an overlap between the two skill groups (Fig. 5). The dataset was separated again into 75% training data and 25 testing data, participants were randomly placed in one of the sets for each round of training. Both models were trained 50 times. KNN provided an average accuracy of 73.3%, while SVC provided 61.0% accuracy. No significant differences in accuracy were found for different values of K.

4   Discussion

In this exploratory data analysis study, we examined the possibility of using supervised classification techniques for automatic labeling of strikes based on trajectory data. Both KNN and SVC were found to accurately classify which limb and technique were used for a strike. KNN was able to classify skill level of the participant with 73.3% accuracy, which indicates that some classification of skill is possible but more data likely needs to be included for accurate classification. Hierarchical clustering was found to increase the accuracy of technique classifications. Considering the excellent accuracy of limb classification, it seems

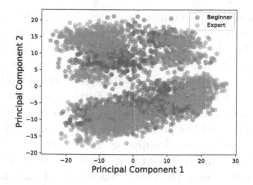

Fig. 5. Principal components colored by skill level. (Color figure online)

recommended to classify strike trajectories on a limb level first and then proceed to the classification of techniques for optimal accuracy. Furthermore, only a small fraction of the data was needed as training data in order to obtain accurately performing models.

4.1 Applications

Automatic classification of strike trajectories allows for rapid analysis of large strike datasets. Until now, manual classification was required in order to process the data obtained in experiments involving the punching bag. This exploratory study is a step towards the (partial) automation of the data analysis. For subsequent experiments with similar set ups, the models developed in this article will be usable to classify the majority of strikes automatically. Furthermore, the small fraction of data that is required to make accurate predictions suggests that this method is also applicable to other experiments for which the current model cannot be applied directly. This would require labeling a small part of the data and then using that to classify the rest. Besides automating data analysis in experiments, this work also contributes to developing a quantitative measure of what distinguishes a strike performed by an expert compared to a novice. The possibility of classifying trajectories based on the skill level of the person performing the strike indicates that strike trajectories have properties that are correlated to the quality of the strike. Although classification of the principal components of the trajectories do not directly give an insight into which aspects of the trajectory predict skill level, this finding implies the possibility of quantifying the quality of a strike based on its trajectory. Accuracy of skill prediction can likely be improved further by including measures such as the impact force and execution time of the strike.

4.2 Limitations and Further Developments

The models presented here still face several limitations. Due to the requirement of knowing the labels beforehand, the models are incapable of detecting techniques that have not previously occurred. This limits the applicability of supervised classification in situations where rare techniques are relevant to the research question. This is the case for research on movement creativity and variability [9]. Therefore, the method should be refined further in order to be able to classify rare techniques. Similarly, accurate prediction of skill will require additional features to be added into the models. Furthermore, the model can only be generalized to datasets that are similar to the dataset used here. Trajectory data that differs significantly will require at least partial manual labeling in order to apply supervised classification. Future research could test the generalizability of the model by attempting to classify previously unseen strike trajectory data. Furthermore, it could include the use of unsupervised classification methods in order to further reduce the need for manual labeling and improve the classification accuracy. Overall, the application of machine learning to the automatic classification of strike data seems like a promising opportunity to advance data analysis in the field of movement science.

References

1. Ashker, S.E.: Technical and tactical aspects that differentiate winning and losing performances in boxing. Int. J. Perform. Anal. Sport **11**(2), 356–364 (2011)
2. Cover, T., Hart, P.: Nearest neighbor pattern classification. IEEE Trans. Inf. Theory **13**(1), 21–27 (1967)
3. Hassanat, A.B., Abbadi, M.A., Altarawneh, G.A., Alhasanat, A.A.: Solving the problem of the K parameter in the KNN classifier using an ensemble learning approach. arXiv preprint arXiv:1409.0919 (2014)
4. Herault, R., Orth, D., Seifert, L., Boulanger, J., Lee, J.A.: Comparing dynamics of fluency and inter-limb coordination in climbing activities using multi-scale Jensen-Shannon embedding and clustering. Data Mining Knowl. Discov. **31**(6), 1758–1792 (2017)
5. Hölbling, D., Preuschl, E., Hassmann, M., Baca, A.: Kinematic analysis of the double side kick in pointfighting, kickboxing. J. Sports Sci. **35**(4), 317–324 (2017)
6. Hsu, C.W., Lin, C.J.: A comparison of methods for multiclass support vector machines. IEEE Trans. Neural Netw. **13**(2), 415–425 (2002)
7. Kasiri, S., Fookes, C., Sridharan, S., Morgan, S.: Fine-grained action recognition of boxing punches from depth imagery. Comput. Vis. Image Underst. **159**, 143–153 (2017)
8. Neto, O.P., Magini, M., Saba, M.M., Pacheco, M.T.T.: Comparison of force, power, and striking efficiency for a Kung Fu strike performed by novice and experienced practitioners: preliminary analysis. Percept. Mot. Skills **106**(1), 188–196 (2008)
9. Orth, D., van der Kamp, J., Memmert, D., Savelsbergh, G.J.: Creative motor actions as emerging from movement variability. Front. Psychol. **8**, 1903 (2017)
10. Pedregosa, F., et al.: Scikit-learn: machine learning in python. J. Mach. Learn. Res. **12**(Oct), 2825–2830 (2011)
11. Song, Y., Huang, J., Zhou, D., Zha, H., Giles, C.L.: IKNN: informative k-nearest neighbor pattern classification. In: Kok, J.N., Koronacki, J., Lopez de Mantaras, R., Matwin, S., Mladenič, D., Skowron, A. (eds.) PKDD 2007. LNCS (LNAI), vol. 4702, pp. 248–264. Springer, Heidelberg (2007). https://doi.org/10.1007/978-3-540-74976-9_25

12. Weston, J., Watkins, C.: Multi-class support vector machines. Technical report. Citeseer (1998)
13. Whiting, W.C., Gregor, R.J., Finerman, G.A.: Kinematic analysis of human upper extremity movements in boxing. Am. J. Sports Med. **16**(2), 130–136 (1988)
14. Zatsiorsky, V.M., Zaciorskij, V.M.: Kinetics of Human Motion. Human Kinetics, Champaign (2002)

Challenge Papers

Predicting Pass Receiver in Football Using Distance Based Features

Yann Dauxais and Clément Gautrais[(✉)]

Univ. Rennes, INRIA, INSA, CNRS, IRISA, Rennes, France
clement.gautrais@irisa.fr

Abstract. This paper presents our approach to the football pass prediction challenge of the Machine Learning and Data Mining for Sport Analytics workshop at ECML/PKDD 2018. Our solution uses distance based features to predict the receiver of a pass. We show that our model is able to improve prediction results obtained on a similar dataset. One particularity of our approach is the use of failed passes to improve the predictions.

Keywords: Pass prediction · Distance based features ·
Interception prediction

1 Introduction

Thanks to the availability of football in-game events, recent approaches have studied aspects of the game that provide actionable knowledge for football clubs. For example, passing patterns from La Liga teams have been studied [8]. Other work focused on estimating football players performance from their actions on the field [1].

The football pass prediction challenge of the Machine Learning and Data Mining for Sport Analytics workshop at ECML/PKDD 2018 focuses on an aspect of the game that has been seldom studied. Indeed, few work have studied the problem of predicting the receiver of a pass in football. In [7], Vercruyssen et al. study the performance of predicting the receiver of a pass, based on different types of features: static or dynamic, quantitative or qualitative and relational or non-relational. They recommend using static qualitative features, such as coned based calculus [2] and double cross calculus [9]. This study is very relevant to our work, as the dataset they are studying is very similar to ours.

In [5], Steiner studies the effects of perceptual information on the probability for a player to receive a pass. He concludes that players with open passing lanes have a higher chance of receiving a pass. Players who are located forward of the passer, close to him or far from an opponent also have a higher chance of receiving a pass. This study is based on quantitative features, and shows pass prediction results similar to [7]. It should be noted that the data used in [5] is not based on real game data, but on scenarios that are evaluated by football

U. Brefeld et al. (Eds.): MLSA 2018, LNAI 11330, pp. 145–151, 2019.
https://doi.org/10.1007/978-3-030-17274-9_12

players in an offline setting. This might have an influence on the pass decision process of the player.

Work have also been dedicated to the prediction of whether a pass will fail [3,6], that is whether the opposing team will intercept the ball during the pass. Important factors are ball velocity, defenders distance to the ball carrier [6] and distances between players [3].

2   Dataset Presentation

The dataset $\mathcal{D}$ contains 12,124 passes performed during 14 different games involving a Belgian football club during the 2014/2015 football season[1]. For each pass, we have the x and y positions of all players, as well as the sender and receiver of the pass. We also have the number of seconds elapsed since the beginning of the half when the pass is performed and received. This yields a total of 60 columns ($2 * 28$ positions, 2 players id and 2 number of seconds). Players 1 to 14 belong to the home team; players 15 to 28 belong to the away team. A thorough description of the dataset can be found on the challenge repository (see footnote 1).

Formally, we have $\mathcal{D} = \langle pass_1, \ldots, pass_j, \ldots, pass_{12124} \rangle$, with $pass_j = \langle x_1, y_1, \ldots, x_i, y_i, \ldots, x_{28}, y_{28}, i_s, i_r, t_s, t_r \rangle$ the j-th pass. x_i and y_i are the coordinates of player i, p_s is the pass sender id, p_r the pass receiver id, t_s the pass sending time and t_r the receiving time.

2.1   Dataset Exploration

We here briefly present some of the characteristics of the dataset. First, we have the positions of 28 players, but there are only 11 players on the field for each team. Fortunately, the substitutes are easily identified, as their x and y positions correspond to the Not a Number (NaN) token. Players having at least one coordinate equal to NaN are removed from the pass vector $pass_j$.

This process yields a total of 22 x and y positions in $pass_j$: one for each player on the field. However, it can happen that a team has less than 11 players on the field, because of an injury or a red card for example. There are 438 pass examples (3.6% of the dataset) were less than 22 players are on the field.

In [7], the authors consider successful passes only: passes where the receiver and sender are on the same team. In the dataset, there are 2077 examples of failed passes (17.1% of the dataset). Because these passes represent a significant portion of the dataset, we decided to keep them.

We removed passes where the sender and receiver are the same players (6 cases), and passes were either the sender or the receiver have NaN positions (2 cases). The final dataset contains 12116 passes.

---

[1] https://github.com/JanVanHaaren/mlsa18-pass-prediction.

3   Model Description

The problem definition of the pass prediction challenge is quite simple.
Challenge problem: For each pass, given the sender and the positions of players, predict the receiver.

3.1   General Approach

The idea of our approach is to estimate the probability P_i that the sender (player p_s) passes to player i, $i \neq p_s$. Then, sorting players id in decreasing order of probability P_i yields a ranking on the pass receiver prediction. Therefore, the player with rank 1 is the player with index equal to $arg\_max_{i \in [1,22], i \neq p_s} P_i$. We now present the learning of the probabilities P_i.

3.2   Learning Pass Probabilities

To learn the pass probabilities P_i, we train a classifier, that, given a set of features, outputs the probability that player i receives the pass from player p_s. Previous approaches predicting the receiver of a pass use different types of features. While the use of qualitative features (relative positioning of players, with respect to the sender and the receiver of the pass) is recommended [7], methods based on quantitative features (distances between players and pass lanes) yield similar results in the prediction of the pass receiver [5].

To train our classifiers, we use both quantitative and qualitative features. Given the position of the sender and of player i (the potential receiver), we first compute different distance based features. First, we compute the Euclidean distance between both players. Afterwards, we compute the forward distance of the pass, that the difference between player i $x$coordinate and the sender x coordinate. Then, we compute the smallest distance between an opponent of the sender and the pass lane. This corresponds to the smallest distance between the line formed by the sender and potential receiver and an opponent of the sender. Next, we compute the smallest and second smallest distance between the sender and players of the opposing team. We also compute the smallest and second smallest distance between the player i and players of the opposing team. Then, we compute a boolean indicating whether the potential receiver and the pass sender are in the same team. Finally, we encode the position of the sender and of player i using a regular grid over the football field. We split the field into 6 rows and 9 columns. Then, the position of a player corresponds to its grid number.

To summarize, we have 10 features to estimate the probability P_i that the sender passes to player i: the distance between the sender and player i; the forward distance of the pass; the smallest distance between the pass line and an opponent; the smallest and second smallest distance between the sender and an opponent; the smallest and second smallest distance between player i and an opponent; the sender and receiver grid number and whether the sender and player i are in the same team. With these 10 features, the classifier predicts whether this pass is the one chosen by the sender. If the sender chose to pass to player i, the value to predict is 1, and 0 otherwise.

4 Results

This section presents the pass receiver prediction results. This corresponds to evaluating the ranking yielded by ordering players id in descending value of P_i. In the followings, this ordered list of players id is called V and v_j refers to the player id at the jth position in this list V.[2]

Table 1. MRR and top-1, top-2 and top-3 recall for predicting the receiver id of a pass. The three lines correspond to the whole pass dataset, the successful pass dataset and the failed pass dataset respectively.

Pass type	MRR	top-1	top-2	top-3
All	0.886($\pm$0.005)	0.841($\pm$0.006)	0.890($\pm$0.006)	0.915($\pm$0.006)
Successful	0.909($\pm$0.006)	0.861($\pm$0.006)	0.919($\pm$0.004)	0.947($\pm$0.003)
Failed	0.779($\pm$0.016)	0.746($\pm$0.018)	0.752($\pm$0.019)	0.763($\pm$0.016)

We train different classifiers (random forest, logistic regression and k nearest neighbors) to predict P_i. We found that the best performance was achieved when using a random forest with 200 trees. To evaluate the performance of the ranking, we use the same metrics as [7]: the mean reciprocal rank measure (MRR): $MRR = \frac{1}{n} \sum_{i=1}^{n} \frac{1}{rank}$, with n the number of examples, $rank$ the rank of the receiver ($rank \in [1, 21]$), and the recall in the top-1, top-2 and top-3 of the ranking. These recalls are equivalent to recall at k [4] for k set up to 1, 2 and 3 respectively: $R@k(V) = \dfrac{\sum_{j=1}^{k} rel(v_j)}{\sum_{j=1}^{m} rel(v_j)}$ where m is the length of V and $rel(v_j)$ is the relevance of v_j and $rel(v_j) = 1$ if v_j is the receiver id and 0 otherwise. It is worth noticing that one and only one player id is relevant for each pass what means that $\sum_{j=1}^{m} rel(v_j) = 1$ and $R@k(V) = \sum_{j=1}^{k} rel(v_j)$. This recall is then averaged on the pass set like for MRR: $\frac{1}{n} \sum_{i=1}^{n} R@k(V)$. Thereby, the top-1 recalls in the followings are equivalent to accuracy. Reported results correspond to the mean value of the corresponding metric on a 5-folds cross-validation. Table 1 presents the result for these 4 measures.

---

[2] Our code is at https://gitlab.inria.fr/cgautrai/prediction_mlsa2018.git.

As one can see, the classifier predicts the good receiver from a list of 21 players in than 84.1% cases. This good classification result is improved to 91.5% when predicting the 3 most probable receivers. We can see by comparing the second and third rows of the table that it is easier to predict the receiver of a successful pass than of an intercepted pass. Furthermore, the prediction improvement between top-1 and top-3 for successful pass is 8.6% when the same improvement for intercepted pass is only 1.7%. It shows that the error on the receiver of an intercepted pass is much higher than for a successful pass.

Table 2. MRR and top-1, top-2 and top-3 recall for predicting the receiver id of a successful pass. The first line corresponds to the results obtained in [7] and the 2 other ones to ours: Random Forest (RF) and Decision Tree (DT).

Method	MRR	top-1	top-2	top-3
Best from [7]	0.42	0.279	0.416	0.467
Our best (RF)	0.870(±0.006)	0.803(±0.009)	0.877(±0.005)	0.922(±0.005)
DT 555 leaves	0.664(±0.004)	0.507(±0.005)	0.675(±0.005)	0.777(±0.007)

To compare our results with the ones obtained in [7], we restrict ourselves to successful passes only, that is passes between two players of the same team. In this case, the *rank* used for the MRR computation belongs to the interval $[1, 10]$. The results are presented in Table 2. It is clear that our approach outperforms the results obtained in [7] on a similar dataset. Our MRR is 2 times better and our top-1 recall 3 times better. It should be noted that the best model in [7] uses 555 rules. To perform a more relevant comparison, we show the results for a decision tree having 555 leaves. This classifier still outperforms the results presented in [7], showing the relevance of our approach.

4.1   Learning from Failures

One of the differences between our approach and the one of Vercruyssen et al. [7] is that we consider opponents as potential receivers. While predicting who will intercept a failed pass is more difficult than predicting who will receive a successful pass, learning from both failed and successful pass can help correct some successful pass decisions. Indeed, when learning from successful passes only, one might over-estimate the probability of a dangerous pass to be successful. Using failed passes can help the classifier to identify these dangerous passes more accurately.

This effect can be observed from the analysis of Tables 1 and 2. Indeed, the MRR for predicting successful passes when using a classifier trained on both successful and failed passes is 0.909 (second line of Table 1); whereas the MRR for predicting successful passes when using the same classifier trained on successful passes only is 0.870 (second line of Table 2). One interesting thing to note is that the top-1 recall difference between both methods is of 5.8%, while the top-3 recall difference is of 2.5%. This means that while both approaches have a similar top 3 ranking for passes, training on both successful and failed passes leads to a better top 1 ranking. It is also worth noting that, for all passes, the MRR lower bound is equal to $\frac{1}{21}$ whereas, for successful passes only, this lower bound is $\frac{1}{10}$. Thereby, it is unfair to compare the MRR of all passes to the one of successful passes only.

4.2 Qualitative Analysis

We now quickly analyze some predictions of our method by looking at 3 pass examples. These passes are represented in Fig. 1.

The top-left example shows a case where our method correctly identifies the receiver: the label 1 is below the orange sign. The receivers ranked 2 and 3 are also teammates of the sender. The rank 2 player is unlikely to receive the ball, as he is close to 2 opponents. The goalkeeper, ranked 3, is however a safe pass option. We can see that our model suggests good passing options, even though the rank 2 player does not seem appropriate.

The top-right image also shows a case where our method correctly identifies the receiver. One interesting thing to note in that example is that the player ranked 2 is actually an opponent of the sender. An even more interesting fact is that this player is likely to intercept the pass between the sender and the actual receiver. This shows that our method is capable of putting high ranks for interceptions, if they make sense. Finally, the bottom example shows an

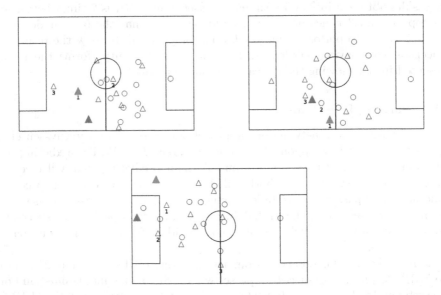

Fig. 1. Top 3 ranking for pass receivers in 3 cases. The pass sender is in green, the pass receiver in orange, the sender teammates in blue and the sender opponents in purple. The ranking is indicated as a red number below the symbol. (Color figure online)

case where our method is not able to predict the receiver in the top 3. While all options chosen by our method seem reasonable, the sender decided to choose another option. Overall, we can say that our method is able to output meaningful rankings, with some exceptions. These exceptions are mostly due to the fact that our method only takes into account distances. Adding new features can help to further eliminate unlikely passes.

5   Conclusion

In this paper, we have illustrated how a simple method, using distance based features, is able to accurately identify the receiver of a football pass. This is done by computing the probability to receive the pass for each player and by predicting the potential receiver with the highest probability. Such approach allows to use a biggest feature set, an thereby, can be generalized. The particularity of this approach is that it also considers failed passes, and predicts the opponent intercepting the pass. For future work, we aim to add new features to strengthen the prediction of the receiver. A good starting point would be to add qualitative features described in [7].

References

1. Decroos, T., Van Haaren, J., Dzyuba, V., Davis, J.: STARSS: a spatio-temporal action rating system for soccer. In: Machine Learning and Data Mining for Sports Analytics ECML/PKDD 2017 Workshop (2017)
2. Frank, A.U.: Qualitative spatial reasoning with cardinal directions. In: Kaindl, H. (ed.) 7. Österreichische Artificial-Intelligence-Tagung/Seventh Austrian Conference on Artificial Intelligence. INFORMATIK, vol. 287, pp. 157–167. Springer, Heidelberg (1991). https://doi.org/10.1007/978-3-642-46752-3_17
3. Horton, M., Gudmundsson, J., Chawla, S., Estephan, J.: Automated classification of passing in football. In: Cao, T., Lim, E.-P., Zhou, Z.-H., Ho, T.-B., Cheung, D., Motoda, H. (eds.) PAKDD 2015. LNCS, vol. 9078, pp. 319–330. Springer, Cham (2015). https://doi.org/10.1007/978-3-319-18032-8_25
4. McSherry, F., Najork, M.: Computing information retrieval performance measures efficiently in the presence of tied scores. In: Macdonald, C., Ounis, I., Plachouras, V., Ruthven, I., White, R.W. (eds.) ECIR 2008. LNCS, vol. 4956, pp. 414–421. Springer, Heidelberg (2008). https://doi.org/10.1007/978-3-540-78646-7_38
5. Steiner, S.: Passing decisions in football: Introducing an empirical approach to estimating the effects of perceptual information and associative knowledge. Front. psychol. **9**, 361 (2018)
6. Travassos, B., Araújo, D., Davids, K., Esteves, P.T., Fernandes, O.: Improving passing actions in team sports by developing interpersonal interactions between players. Int. J. Sports Sci. Coach. **7**(4), 677–688 (2012)
7. Vercruyssen, V., De Raedt, L., Davis, J.: Qualitative spatial reasoning for soccer pass prediction (2016)
8. Wang, Q., Zhu, H., Hu, W., Shen, Z., Yao, Y.: Discerning tactical patterns for professional soccer teams: an enhanced topic model with applications. In: Proceedings of the 21th ACM SIGKDD ICKDM, pp. 2197–2206. ACM (2015)
9. Zimmermann, K., Freksa, C.: Qualitative spatial reasoning using orientation, distance, and path knowledge. Appl. intell. **6**(1), 49–58 (1996)

Football Pass Prediction
Using Player Locations

Philippe Fournier-Viger[1]([⊠]), Tianbiao Liu[2], and Jerry Chun-Wei Lin[3]

[1] School of Natural Sciences and Humanities,
Harbin Institute of Technology (Shenzhen), Shenzhen, China
philfv8@yahoo.com
[2] College of Sports and Physical Education,
Beijing Normal University, Beijing, China
ltb@bnu.edu.cn
[3] Department of Computing, Mathematics and Physics,
Western Norway University of Applied Sciences (HVL), Bergen, Norway
jerrylin@ieee.org

Abstract. In many sports, predicting the passing behavior of players is desirable at it provides insights that can help to understand and improve player performance. In this paper, we describe a novel model for football pass prediction, developed to participate in the Prediction Challenge of the 5th Workshop on Machine Learning and Data Mining for Sports Analytics, collocated with ECML PAKDD 2018. The model called Football Pass Predictor (FPP) considers various aspects to generate predictions such as the distance between players, the proximity of players from the opposite team, and the direction of each pass. Experimental results shows that the model can achieve a prediction accuracy of 33.8%, and more than 50% if two guesses are allowed. This is considerably more than the random predictor, which obtains 8.3%.

Keywords: Football · Pass prediction · Prediction challenge

1 Introduction

Passing is a key aspect of the football game. It is performed between players, occurs almost everywhere on the pitch, and creates scoring opportunities. According to statistics from the Union of European Football Association, during the 2017–2018 season, a top European team could perform more than 400 passes in a Champions league match [1]. Passing can account for the majority of tactical/technical behaviors in a game. Many researchers have studied passing behavior from the perspective of football tactics and techniques [2,3]. Nowadays, with the development of data science and computer technology, researchers can analyze and simulate passing using massive databases and complicated models, to acquire a deeper understanding of passing behaviors. For instance, Liu et al. [4,5] evaluated the effect of passing on creating scoring opportunities using

© Springer Nature Switzerland AG 2019
U. Brefeld et al. (Eds.): MLSA 2018, LNAI 11330, pp. 152–158, 2019.
https://doi.org/10.1007/978-3-030-17274-9_13

a Markov chain model. In another study, an Apriori-based algorithm was applied to perform a descriptive analysis of passing behavior [4, 5] Apriori-based diagnostical analysis of football passes was also done using a sequential rule discovery approach [6] based on the RuleGrowth algorithm [7].

Passing behavior is influenced by many factors and a player must quickly react to the sudden changes of situations around him. Considering the playing context and different playing situations, Stöckl et al. [10] provided an approach to describe the tactical difficulty of passes. In another study, Rein et al. [11] assessed passing effectiveness in elite soccer using two algorithms considering the number of defenders and players' control of space. Gyarmati et al. [12] developed a QPass evaluation system to estimate players role in building up an attack. Another way to identify key players in a team is using network analysis and to consider pass difficulty [13]. Generally, passing differs according to the context. Sometimes, passings yield dangerous situations with respect to opponents, while sometimes few risks are involved. Cakmak et al. [14] developed a descriptive model to quantify the effectiveness of passes and identify key passes and regular passers in a team. Using Player Trajectory technology, Lida and Mase [15] studied the ball passing behavior of players by considering their trajectories. Dhar and Singh [16] analyzed video footage to develop a passing strategy. Although, all these studies have been done to analyze football passing behavior, they do not provide a computational model that can be used for pass prediction. But pass prediction could have many applications.

The contribution of this paper is to fill this void by proposing a model named FPP (Football Pass Predictor) to predict the player who will receive a pass initiated by another player. This work is done in the context of the Prediction Challenge of the 5th Workshop on Machine Learning and Data Mining for Sports Analytics, collocated with ECML PAKDD 2018. The proposed model considers various aspects to generate predictions such as the distance between players, the proximity of players from the opposite team and the direction of each pass. Experimental results shows that the model can achieve high prediction accuracy.

The rest of this paper is organized as follows. Section 2 provides a brief description of the provided dataset, and key observations that were made. Section 3 presents the proposed FPP model. Section 4 presents experimental results. Finally, Sect. 5 draws a conclusion.

2   Observations About the Data

The dataset provided for the prediction challenge contains 12,124 records describing passes from 14 football matches of a Belgian team and opposing teams during the 2014/2015 football season. In a football match, two teams are facing each others, where each team has 14 players (including 3 substitutes). A database record describes a pass. It provides the (1) the location of the 14 football players of each team using 2D coordinates, (2) the time at which the pass started and ended, (3) the player who sent the ball, and (4) the player who received the ball. Coordinates are expressed in the $[-5250, 5250]$ $[-3400, 3400]$

intervals for the X and Y axes, respectively. Note that player names are not indicated in the data as well as the names of the teams. Moreover, it has not been indicated if the positions of the players have been recorded when a pass starts or ends. Besides, although timestamps are provided in the data, records from all matches were put in a single file and randomly shuffled. Thus, each pass can only be considered individually rather than in the context of a match. The data was collected by the prediction challenge organizers, and made available at https://github.com/JanVanHaaren/mlsa18-pass-prediction.

By analyzing the data, the authors of this paper made a few interesting observations. First, out of the 12,124 passes, only 17% of the passes are intercepted by the opposite team. Because unsuccessful passes are much less likely than successful ones, a design principle for the proposed model is to assume that all passes will be successful when making predictions. Second, if was found that the 163th line of the dataset is an invalid record. In that record, the player number 15 who sends the ball has no coordinates. This record has been ignored. Third, although the dataset provides timestamps, it is difficult to use this information for pass prediction since each record is often separated by numerous seconds, and the position of players is given only once for each pass but each record contains two timestamps. For this reason, the trajectories of players are not available, and it is hence difficult to analyze each pass in the context of the overall game. Fourth, a related issue is that records from all matches are stored together in the dataset. Thus, it is unclear which passes belong to which match. Besides, it is not indicated which team is playing on which side of the football field. We have inferred this information by assuming that the left (right) side of the pitch belongs to the team having the leftmost (rightmost) player.

3   The FPP Model

Based on the observations made on the data, the proposed FPP model was developed. To design the model an iterative design approach was used where several versions of the model were successively designed, each adding additional criteria to increase prediction accuracy. Among the multiples versions of the model, four are described in the paper. These versions, sorted by ascending order of complexity, are called M1, M2, M3, and FPP, respectively. They are described in the following paragraphs, and illustrated in Fig. 1.

M1. The first model is based on the assumption that the sender will pass the ball to the closest player of his team. Assume that a player X has the ball and that we want to predict who will receive the ball. Let P be the set of players from the same team as X (excluding X). For each player $Y \in P$, the Euclidian distance between X and Y is calculated, denoted as $d_{X,Y}$. Then, for each player $Y \in P$ a score is assigned to Y, defined as $score(Y) = d(X,Y)$. The player with the smallest score is chosen as the prediction.

M2. The second model is an improvement of the M1 model. An additional idea is considered, which is that a player may be less likely to receive the ball if a player of the opposite team is close to him. The motivation is that this situation may be considered more risky for the sender, and that the opposite team player may intercept the ball. Formally, let O be the set of players from the opposite team. For each player $Y \in P$, its score is defined as $score(Y) = d(X,Y) + penaltyC(Y,O)$. The term $penaltyC$ is defined as $penaltyC(Y,O) = 900$ if there exists a player $Z \in O$ such that $d(Y,Z) < 700$, and otherwise $penaltyC(Y,O) = 0$. The values 700 and 900 were found empirically (by trial and error) to obtain a high prediction accuracy.

M3. The third model is an improvement of M2, which considers that more than one player from the opposite team may be close to a potential receiver and increase risks. For each player $Y \in P$, its score is defined as $score(Y) = d(X,Y) + penaltyC(Y,O) + penaltyD(Y,O)$. The term $penaltyD$ is defined as $penaltyD(Y,O) = 55$ if there exists two players $Z \in O$ such that $d(Y,Z) < 700$, and otherwise $penaltyD(Y,O) = 0$. The value 55 was found empirically.

FPP. The fourth model is an improvement of the M3 model, which considers the direction of the ball, based on the assumption that a player prefers to send the ball forward. Let the notation $Z.x$ denotes the position of a player Z on the x axis. The score of a player $Y \in P$ is defined as $score(Y) = d(X,Y) + penaltyC(Y,O) + penaltyD(Y,O) + direction(X,Y)$. The term $direction(X,Y)$ is defined as $-0.3 \times |X.x - Y.x|$ if the pass is a forward pass (toward the opposite team goal) or as $0.1 \times |X.x - Y.x|$ if the pass is a backward pass. The values 0.1 and 0.3 were found empirically as providing the best results.

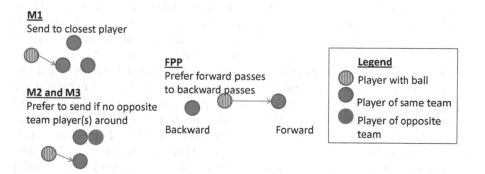

Fig. 1. An illustration of the ideas introduced in the proposed models

4   Experimental Evaluation

An experimental evaluation was performed to evaluate the five versions of the designed FPP model. The models were compared with a random predictor as baseline. Since no performance measure was explicitly specified for the prediction challenge, it was decided to evaluate the models in terms of accuracy (number of correct predictions divided by total number of passes to be predicted). Furthermore, the accuracy when two guesses are allowed was measured. In that situation, a model can make two predictions for each pass, and if one of them is right, the pass is considered as correctly predicted. The source code of the proposed model and evaluation framework can be downloaded from http://philippe-fournier-viger.com/foot2018/. It is written in Java.

Table 1. Accuracy of the compared models

Model	Accuracy for one guess (%)	Accuracy for two guesses (%)
Baseline (random)	7.88	16.51
M1	27.44	46.58
M2	32.83	50.92
M3	32.97	50.94
FPP	**33.38**	**51.81**

Results are shown in Table 1. It is first observed that the heuristic of predicting that the ball is passed to the closest player of the same team achieves a high accuracy (27.44%) compared to the baseline random predictor (7.88%). Then, if we add a penalty if there is a player from the opposite team that is close to a potential receiver, it increases accuracy by more than 5%, from 27.44% (M1) to 32.83% (M2). If this model is further extended for the case of two players from the opposite team, the accuracy increases slightly, from 32.83% (M2) to 32.97% (M3). Moreover, if the direction of the ball is considered, the accuracy increases from 32.97% (M4) to 33.38% (FPP). Finally, if we allows to perform two guesses, the accuracy of the best model (FPP) increases to 51.81%.

Besides, the described models, the authors of this paper have also tried various other ideas including calculating the angle between players to determine if a player from the opposite team may intercept a pass. But these ideas did not improve accuracy, or even decreased it. Other models could also be considered. However, what can be developed remains limited by the data. For example, having more rich data such as the real time locations of players, and data were records are not shuffled, could allow to obtain player trajectories and develop more complex models. Besides, an improvement of the FPP model could be to use a genetic algorithm to tune its parameters instead of tuning them by hand, and to split the data in training and testing sets, or using k-fold cross validations to avoid the potential problem of overfitting.

5   Conclusion

This paper has proposed a model called Football Pass Predictor (FPP) to predict the receivers of passes in football matches. The model considers various aspects such as the distance between players, the proximity of players from the opposite team, the direction of each pass, to generate predictions. The performance of the model was compared with a baseline random predictor and several variations of the proposed models. Results from experiments shows that FPP can achieve a prediction accuracy of 33.8%, and more than 50% if two guesses are allowed. An interesting perspective for future work is to collect richer data, which would allow to develop more complex models. We also plan to evaluate the possibility of using pattern mining approaches for football pass prediction [8,9].

Acknowledgement. This work is partially financed by the Youth 1000 talent funding of Philippe Fournier-Viger.

References

1. http://www.uefa.com/uefachampionsleague/season=2018/statistics/round=2000881/matches/kind=passes/index.html . Accessed 15 June 2018
2. Garratt, K., Murphy, A., Bower, R.: Passing and goal scoring characteristics in Australian A-League football. Int. J. Perform. Anal. Sport **17**(1–2), 77–85 (2017)
3. Plummer, B.T.: Analysis of attacking possessions leading to a goal attempt, and goal scoring patterns within men's elite soccer. J. Sports Sci. Med. **1**(1), 1–38 (2013)
4. Liu, T.: Systematische Spielbeobachtung im internationalen Leistungsfußball. Ph.D. dissertation. University of Bayreuth (2014)
5. Liu, T., Hohmann, A.: Apriori-based diagnostical analysis of passings in the football game. In: Proceedings of IEEE 2016 International Conference on Big Data Analysis, pp. 1–4. IEEE (2016)
6. Liu, T., Fournier-Viger, P., Hohmann, A.: Using diagnostic analysis to discover offensive patterns in a football game. In: Tavana, M., Patnaik, S. (eds.) Recent Developments in Data Science and Business Analytics. SPBE, pp. 381–386. Springer, Cham (2018). https://doi.org/10.1007/978-3-319-72745-5_43
7. Fournier-Viger, P., Nkambou, R., Tseng, S.M.: RuleGrowth: mining sequential rules common to several sequences by pattern-growth. In: Proceedings of 26th Symposium on Applied Computing, pp. 954–959. ACM Press (2011)
8. Fournier-Viger, P., Lin, J.C.-W., Vo, B., Chi, T.T., Zhang, J., Le, H.B.: A survey of itemset mining. WIREs Data Min. Knowl. Discov. e1207 (2017). https://doi.org/10.1002/widm.1207
9. Fournier-Viger, P., Lin, J.C.-W., Kiran, R.U., Koh, Y.S., Thomas, R.: A survey of sequential pattern mining. Data Sci. Pattern Recogn. (DSPR) **1**(1), 54–77 (2017)
10. Stöckl, M., Cruz, D., Duarte, R.: Modelling the tactical difficulty of passes in soccer. In: Chung, P., Soltoggio, A., Dawson, C.W., Meng, Q., Pain, M. (eds.) Proceedings of the 10th International Symposium on Computer Science in Sports (ISCSS). AISC, vol. 392, pp. 139–143. Springer, Cham (2016). https://doi.org/10.1007/978-3-319-24560-7_17
11. Rein, R., Raabe, D., Memmert, D.: "Which pass is better?" Novel approaches to assess passing effectiveness in elite soccer. Hum. Mov. Sci. **55**, 172–181 (2017)
12. Gyarmati, L., Stanojevic, R.: QPass: a merit-based evaluation of soccer passes. Preprint on arXiv:1608.03532 (2016)
13. McHale, I.G., Relton, S.D.: Identifying key players in soccer teams using network analysis and pass difficulty. Eur. J. Oper. Res. **268**(1), 339–347 (2018)

14. Cakmak, A., Uzun, A., Delibas, E.: Computational modeling of pass effectiveness in soccer. J. Adv. Complex Syst. (2018, in press)
15. Lida, R., Mase, K.: Ball passing course creating behavior in soccer game detection from player trajectory. IEICE Technical report, vol. 113, no. 432, pp. 171–176 (2014)
16. Dhar, J., Singh, A.: Game analysis and prediction of ball positions in a football match from video footages. In: Proceedings of International Conference on Recent Advances and Innovations in Engineering, pp. 1–6. IEEE (2014)

Deep Learning from Spatial Relations for Soccer Pass Prediction

Ondřej Hubáček$^{(\boxtimes)}$, Gustav Šourek, and Filip Železný

Czech Technical University, Prague, Czech Republic
{hubacon2,souregus,zelezny}@fel.cvut.cz

Abstract. We propose a convolutional architecture for learning representations over spatial relations in the game of soccer, with the goal to predict individual passes between players, as a submission to the prediction challenge organized for the 5th Workshop on Machine Learning and Data Mining for Sports Analytics. The goal of the challenge was to predict the receiver of a pass given location of the sender and all other players. From each soccer situation, we extract spatial relations between the players and a few key locations on the field, which are then hierarchically aggregated within the neural architecture designed to extract possibly complex gameplay patterns stemming from these simple relations. The use of convolutions then allows to efficiently capture the various regularities that are inherent to the game. In the experiments, we show very promising performance of the method.

1 Introduction

Predictive sport analytics is a modern discipline where various statistical models are employed to assess different aspects of a game. In this paper, we focus on the game of soccer and predicting individual passes between players during the match, given a static snapshot of each pass situation, i.e. indication of ball possession and locations of all the players. This setting was given by the prediction challenge organized for the 5th Workshop on Machine Learning and Data Mining for Sports Analytics held in conjunction with ECML.

Since each learning example is, in this case, just an independent, static viewpoint on the game, we approach the problem from a simple geometrical perspective. In that view, we take each situation, determined by mere absolute locations of the players, enrich these with a few soccer-specific contextual locations, and turn their absolute positions into mutual, relative distances. This way we enable to the model to generalize across different situations, reasoning about the mutual spatial patterns between the players, rather than their positions on the filed. These spatial patterns are represented with convolutional filters, capturing the inherent symmetries and geometrical regularities arising from the rules of the game. These patterns are then further aggregated with pooling and combined in a fully connected manner to help the model to explore their relations. As opposed to some existing works based quite heavily on expert knowledge, we employ just a very few assumptions on the patterns and rather aim at the benefits of end-to-end learning.

U. Brefeld et al. (Eds.): MLSA 2018, LNAI 11330, pp. 159–166, 2019.
https://doi.org/10.1007/978-3-030-17274-9_14

1.1   Related Work

Inductive logic programming model [7] trained on qualitative spatial representations [2] was previously used to tackle the task of predicting soccer passes. Similar approach was used for discovering offensive patterns [6]. Spatio-temporal data were further utilized to infer teams' play-styles [1,4] and to examine the likelihood of scoring a goal from a shot [3]. Another approach leveraged a physics-based model of soccer ball motion to predict the receiver of the pass [5].

1.2   Dataset

The dataset consisted of 12 124 soccer passes from which 10 045 passes were successful (meaning that the sender and the receiver of the pass were from the same team). We decided to focus on predicting only the successful passes as was done previously [7].

Unlike in the previous work [7], the dataset contained solely the snapshot of the game in form of coordinates of each of the 22 players on the field, making the situations independent of each other. This makes the prediction task much harder, because we have no information about players' momentum, or orientation in space etc. Neither were are capable to determine the same team or player across multiple situations. The dataset also contained the timestamp of when the pass was send and received. Due to the predictive nature of the task, we decided to omit the timestamp of the pass receipt, since it is obviously not available when making the actual prediction.

In 367 cases, only 21 players' coordinates were present, presumably after one player had been sent off. To deal with the missing coordinates we inputted surrogate large numbers as the coordinates, so this position became meaningless for the predictions.

2   Predictive Model

The proposed model is a neural architecture consisting of a convolutional layers with diverse filters, max-pooling and a fully connected layers with a softmax output. Each of the convolutional filters encodes a certain feature-set transformation designed to extract a particular context from the game snapshots. Intuitively, these may collect information on how occupied the potentially receiving player is, how pressured the sender of the pass is, or where is the receiver positioned on the field w.r.t. his teammates. The max-pooling layer helps the model to become agnostic to the particular positioning and ordering of the players in order to generalize better, based on the intuition that typically only a very few closest players are relevant to each pass. The softmax output then naturally encodes the exclusive outcomes of each situation, since only one pass at a time is ever carried out.

Table 1. Enriching spatial snapshots with contextual locations.

Feature	Description
$f_1 : dist(p_s, p_r)$	Distance between sender p_s and potential receiver p_r
$f_2 : dist(p_s, goal)$	Distance of p_s to the center of the opponent's goal
$f_3 : dist(p_s, side)$	Perpendicular distance of p_s to closest sideline
$f_4 : dist(p_s, center_{t_1})$	Distance of p_s to the center of gravity of his teammates
$f_5 : dist(p_s, center_{t_2})$	Distance of p_s to the center of gravity of his opponents
$f_6 : dist(p_r, goal)$	Distance of p_r to the center of the opponent's goal
$f_7 : dist(p_r, side)$	Perpendicular distance of p_r to closest sideline
$f_8 : dist(p_r, center_{t_1})$	Distance of p_r to the center of gravity of his teammates
$f_9 : dist(p_r, center_{t_2})$	Distance of p_r to the center of gravity of his opponents
$f_{10} : dist(p_r, p_i)$	Distance of p_r to a teammate p_i
$f_{11} : dist(p_s, p_i)$	Distance of p_s to a teammate p_i
$f_{12} : dist\_opp(p_r, p_i)$	Distance of p_r to an opponent p_i
$f_{13} : dist\_opp(p_s, p_i)$	Distance of p_s to an opponent p_i

2.1 Knowledge Representation

The raw data come in a simple table format where, for each pass situation during the course of each game, we are given x-y coordinates of the 22 players on the field with an indicator of the sender of the pass p_s, i.e. a tuple of

$$(timestamp, p_{1_x}, p_{1_y}, \ldots, p_{11_x}, p_{11_y}, p_{12_x}, p_{12_y} \ldots, p_{22_x}, p_{22_y}, p_s)$$

For the purpose of pass prediction, we look at each snapshot from the perspective of potential successful passes between the ball-possessing player p_s and all his teammates (potential receivers) p_r, i.e. for each situation we have 10 pairs of players

$$(p_s, p_r), \text{ such that } p_r \in \begin{cases} \{p_1, \ldots, p_{11}\} \setminus p_s, & \text{if } s \in \{1, \ldots, 11\} \\ \{p_{12}, \ldots, p_{22}\} \setminus p_s, & \text{if } s \in \{12, \ldots, 22\} \end{cases}$$

As a preprocessing step, we enrich these pairs with several key static and dynamic locations from the field, upon which we measure distances as described in Table 1. These enriched pairs, representing the potential passes, then constitute our learning examples.

2.2 Neural Architecture

An overview of the neural architecture is displayed in Fig. 1. At the input to the model, the resulting spatial relations described in Sect. 2.1 are being aggregated into sets to form feature maps for the convolutional filters. Particularly, for each potential pass (p_s, p_r), we conform the relations into different filters expressing different viewpoints on the pass, such as *cover* of the receiving player or *pressure* on the sender and *alternatives* available to him, as detailed in Table 2. Each of these feature sets, or filters, may be instantiated multiple times w.r.t. the variables p_i iterating over the opponents of the sender (*cover*, *pressure*) and teammates of the sender (*alternative*). Within the context of each filter, we order the remaining players w.r.t. the f_{10}, f_{12} and f_{13} for *alternative*, *cover* and *pressure*, respectively. This way we enforce ordering on these instantiations, resulting into 1D feature maps upon which the filters operate, as depicted in Fig. 1. Thus despite the *cover* and *pressure* filters operating on the same feature sets, they will result into different feature maps. Also, since all these filters principally share the common static context of where within the field the current situation occurs, described by the features $f_1 \ldots f_9$, we exclude these from the individual filters to merge them later in the model only to prevent redundancy in the feature maps.

Table 2. Conformation of spatial relations into convolutional filters.

Filter	Features	Context
alternative	(f_1, f_{10}, f_{11})	The alternatives available to the sender
cover	(f_1, f_{12}, f_{13})	The occupation of the sender by his opponents
pressure	(f_1, f_{12}, f_{13})	The pressure on the sender from his opponents

The resulting values from these filters are then aggregated via max-pooling. While multiple pools could be connected with a standard overlay to capture the different sub-regions of the distance space, we set a global pool over all instantiations of each single filter, following the intuition that only the closest players are typically relevant, suppressing the potential noise from the rest. To alleviate this somewhat radical assumption, we also employ wider filters to capture couples of the remaining players rather than individuals. This way we may also reason about more complex spatial patterns between the relevant players. These filters of size 3×2 and 3×3 further distinguish the use of *cover* and *pressure*. Finally the pooling helps to neglect the potentially harmful effect of the, to a certain degree ad-hoc, overall ordering.

The patterns extracted with the help of the filters and selected by the pools form an input to the fully connected layers (Fig. 1). The purpose of these layers is to combine all the different patterns into a final value expressing the potential of each individual pass (p_s, p_r). Intuitively, these layers express the logic of decision making the sender p_s is normally going through, incorporating the relational

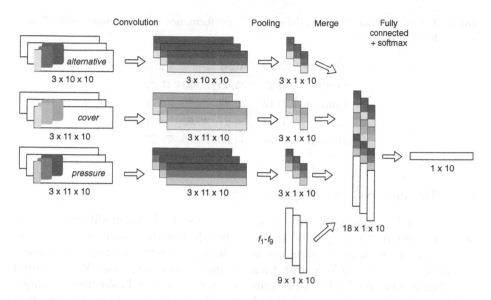

Fig. 1. Architecture of the neural model. Four feature maps of size $\#features \times \#instantiations \times \#possibilities$ are at the input. Filters of size 3×1, 3×2 and 3×3 are applied to each feature map. The outputs of the convolution are reduced by max pooling and merged with the $f_1 - f_9$ features providing their static context. Finally, 2 dense layers with 3, respectively 1, neurons are applied to each possibility. For clarity only 3 out of 10 possibilities (depth dimension) are displayed.

contexts (filters) of the receiving player p_r w.r.t. his own, while learning how to weight the importance of the individual patterns in each combination.

Finally, with the softmax output (Fig. 1), we enable the model to reason jointly over the whole set of 10 possible passes (p_s, p_r). As opposed to separating each pass situation into 10 independent learning examples and normalizing over these as a postprocessing step, with the joint output the gradient directly steers the model towards exclusive predictions as part of the learning process.

3 Experiments

We performed 10-fold crossvalidation, evaluated the model w.r.t. mean reciprocal rank and how many times the actual receiver of the pass was among the three most likely predictions. We compared our result with [7], where the authors made use of both static and dynamic features derived from the flow of the game, which was unavailable to us. Therefore it would be fair to compare our results with the *Static* model from the mentioned work. Nevertheless, our model outperformed both the *Static* and the *Combined* model, which combined static and dynamic features (Table 3).

Table 3. Comparison of the model's (CNN) performance with previous work [7] and human-level performance.

	MRR	top-1	top-2	top-3
Static	0.39	25.49	36.33	41.22
Combined	0.42	27.87	41.59	46.70
CNN	0.48	29.87	45.63	55.91
Human	0.56	34.00	56.25	70.25

3.1 Human-Level Performance

We measured human-level performance to assess the inherent difficulty of the task. We were particularly curious about the effect of the missing dynamic context of the game that humans are used to from standard visual recordings, providing much more information than the mere static snapshots. We measured and averaged the predictive performance of three soccer enthusiasts on a sample of 200 randomly selected situations. To put the data into a more familiar perspective, we created a simple interactive visualization[1] that may be utilized for further measurements. The task proved to be difficult even for humans. While the top-1 accuracies of the model and humans were close, the top-3 accuracies and MMR showed that humans clearly rank the alternatives better.

3.2 Discussion

We analyzed the predictions made by the model to obtain further insights. The main weakness of the model was that it usually considered only a few options as viable, even when their alternatives were very similar. This could be due to the use of softmax in combination with cross-entropy loss when training the network instead of some kind of ranking loss. The network was strong in spotting uncovered teammates, sometimes even overvaluing their positions. Generally, the network preferred passes to sidelines, even when we could guess that the ball most likely just came from those positions. The human intuition thus seems superior when capturing this underlying "flow" of the game.

Visualization of an example situation illustrates the difficulty of the task Fig. 2. Without the information about the senders orientation on the field, there are many viable alternatives. The model marked the pass to the sideline as the most probable as this is a common pattern – midfielder developing a play from the center of the field with a pass to the sideline. The actual pass was the model's second guess. From a human perspective there are far too many options assigned near zero probability. Especially the passes to the players 9 and 2 should have been prioritized more.

---

[1] https://github.com/Hudler/pass-viz.

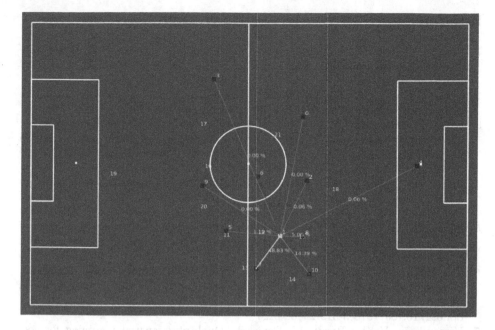

Fig. 2. Example model prediction. Possible passlines are depicted by yellow lines, with the actual pass marked by red. The percentages near the passlines show the predicted probabilities. (Color figure online)

The decomposition of the static context features $f_1 \ldots f_9$ from the convolutional filters, as depicted in Fig. 1, might suggests that the model could be split into two. While these complementary feature sets provide context to each other and were thus meant to work together, we also measured their separate performance, proving the convolutional features to be more valuable (MRR 0.46) than the static context features (MRR 0.42) in separate experiments.

4   Conclusion

We detailed our model for soccer pass prediction given static spatial snapshots of the game. The model was a neural architecture based on a set of convolutional filters, carefully designed to extract different relational contexts from each game situation, i.e. mutual positions of players on the field. We argued how such an architecture may learn possibly complex relational patterns via aggregation of simple spatial relations. Finally, on a large dataset of captured soccer passes, we showed that promising results can be achieved with such an approach.

Acknowledgements. Authors acknowledge support by "Deep Relational Learning" project no. 17-26999S granted by the Czech Science Foundation. Computational resources were provided by the CESNET LM2015042 and the CERIT Scientific Cloud LM2015085, provided under the programme "Projects of Large Research, Development, and Innovations Infrastructures".

References

1. Brooks, J., Kerr, M., Guttag, J.: Using machine learning to draw inferences from pass location data in soccer. Stat. Anal. Data Min. ASA Data Sci. J. 9(5), 338–349 (2016)
2. Chen, J., Cohn, A.G., Liu, D., Wang, S., Ouyang, J., Yu, Q.: A survey of qualitative spatial representations. Knowl. Eng. Rev. 30(1), 106–136 (2015)
3. Lucey, P., Bialkowski, A., Monfort, M., Carr, P., Matthews, I.: Quality vs quantity: improved shot prediction in soccer using strategic features from spatiotemporal data. In: Proceedings 8th Annual MIT Sloan Sports Analytics Conference, pp. 1–9 (2014)
4. Lucey, P., Oliver, D., Carr, P., Roth, J., Matthews, I.: Assessing team strategy using spatiotemporal data. In: Proceedings of the 19th ACM SIGKDD International Conference on Knowledge Discovery and Data Mining, pp. 1366–1374. ACM (2013)
5. Spearman, W., Basye, A., Dick, G., Hotovy, R., Pop, P.: Physics-based modeling of pass probabilities in soccer. In: Proceeding of the 11th MIT Sloan Sports Analytics Conference (2017)
6. Van Haaren, J., Dzyuba, V., Hannosset, S., Davis, J.: Automatically discovering offensive patterns in soccer match data. In: Fromont, E., De Bie, T., van Leeuwen, M. (eds.) IDA 2015. LNCS, vol. 9385, pp. 286–297. Springer, Cham (2015). https://doi.org/10.1007/978-3-319-24465-5_25
7. Vercruyssen, V., De Raedt, L., Davis, J.: Qualitative spatial reasoning for soccer pass prediction. In: Machine Learning and Data Mining for Sports Analytics ECML/PKDD 2016 Workshop (2016)

Predicting the Receivers
of Football Passes

Heng Li[1(✉)] and Zhiying Zhang[2]

[1] School of Computing, Queen's University, Kingston, Canada
hengli@cs.queensu.ca
[2] Microsoft, Bellevue, WA, USA
zhiyingz@microsoft.com

Abstract. Football (or association football) is a highly-collaborative team sport. Passing the ball to the right player is essential for winning a football game. Anticipating the receiver of a pass can help football players build better collaborations and help coaches make informed tactical decisions. In this work, we analyze a public dataset that contains 12,124 passes performed by professional football players. We extract five dimensions of features from the dataset and build a learning to rank model to predict the receiver of a pass. Our model's first, top-3 and top-5 guesses find the correct receiver of a pass with an accuracy of 50%, 84%, and 94%, respectively, when we exclude false passes, which outperforms three baseline models that we use to rank the candidate receivers of a pass. The features that capture the positions of the candidate receivers play the most important roles in explaining the receiver of a pass.

Keywords: Football pass prediction · Learning to rank ·
LambdaMART · Gradient boosting decision tree · LightGBM

1 Introduction

In a football game, players pass the ball to their teammates in order to create good shooting opportunities or prevent the opposing team from getting the control of the ball. Accurately passing the ball to the right player is essential for winning a football game [1,6].

Prior work [6,9] studies how passing sequences lead to goals. Their findings have shaped the tactics of many football coaches. In this work, we build a learning to rank model [8] to anticipate the receiver of a football pass. We believe that football coaches and players can take our results into consideration when they make their tactics or make their passes/runs. For example, a player could learn from the important factors for explaining the receiver of a pass to improve his/her chance of receiving the ball. Anticipating receivers of passes can also help automated cameras always focus on the ball in a game.

© Springer Nature Switzerland AG 2019
U. Brefeld et al. (Eds.): MLSA 2018, LNAI 11330, pp. 167–177, 2019.
https://doi.org/10.1007/978-3-030-17274-9_15

Table 1. A summary of players' passing statistics.

	Back-field	Middle-field	Front-field	Overall
Passing accuracy	86%	83%	79%	83%
Median passing distance (m)	17	14	11	14
Passing forwards ratio	74%	61%	50%	62%

This work analyzes a dataset which contains 12,124 passes performed by a Belgian football club in 14 games[1]. We want to answer the following research questions (RQs):

RQ1: How well can we model the receiver of a pass? We build a learning to rank model to predict the receiver of a football pass. An accurate model can help coaches and players make informed tactical decisions in a game.

RQ2: What are the important factors that explain the receiver of a pass? We analyze the model to find the most influential factors that explain the receiver of a pass. Understanding such influential factors can help coaches and players improve their tactics and passes/runs according to these factors.

Paper organization. The remainder of the paper is organized as follows. Section 2 explores the dataset that we use. Section 3 discusses our approaches for building and evaluating our prediction model. Section 4 presents the results for answering our research questions. Finally, Sect. 5 draws conclusions.

2   Data Exploration

Dataset overview. The dataset contains information about 12,124 football passes. For each pass, the dataset provides the information about the time of the pass since the start of the half, the coordinates of all the players on the pitch, the identifier of the player who passes the ball, and the identifier of the player who receives the ball.

Dealing with missing values. Our goal is to predict the receiver of a pass based on the information about the sender and other players on the pitch (i.e., the candidate receivers). Among the 12,124 passes, there is one pass that misses the coordinates of the sender, and one pass that misses the coordinates of the receiver. For another six passes, the senders and the receivers are the same players. We remove the above-mentioned eight data instances from our dataset. We end up with 12,116 valid passes in our dataset.

Overall, players' passing accuracy is 83%, and the passing accuracy decreases from the back field to the front field. Table 1 shows a summary of players' passing statistics. We define the passing accuracy as the ratio of the

---

[1] https://github.com/JanVanHaaren/mlsa18-pass-prediction.

Table 2. Five-number summary of players' passing distance.

Min.	1st Qu.	Median	3rd Qu.	Max.
0	9	14	20	70

passes that reach a teammate. We divide the field into three equally sized areas along the long side of the field, namely back field, middle field and front field. We define a pass as a **back-field pass, middle-field pass**, or **front-field** pass when the sender is within the back field, the middle field and the front field, respectively.

The median passing distance is 14 m, and the passing distance decreases from the back field to the front field. Table 2 shows the five-number summary of players' passing distance. While the maximum passing distance is 70 m, 75% of the passes are within 20 m. As shown in Table 1, the median passing distance for the back field, middle field, and front field is 17, 14, and 11 m, respectively.

Players pass the ball forwards in 62% of the passes, and the ratio of forward passes decreases from the back field to the front field. In the back field, players pass the ball forwards in 74% of the passes, and the ratio decreases to 61% and 50% for middle-field passes and front-field passes, respectively.

> Players present different passing characteristics in different areas of the field. Such differences suggest us to build and evaluate our model in different areas of the field.

3 Methodology for Predicting the Receivers of Football Passes

This section discusses our overall methodology, including our feature extraction process, modeling and evaluation approaches.

3.1 Feature Extraction

From the dataset that we explain in Sect. 2, we extract five dimensions of features to explain the likelihood of passing the ball to a certain receiver. In total, we extract 54 features. A full list of our features is available at our public github repository[2]. We also share our extracted feature values online[3].

---

[2] https://github.com/henglicad/mlsa18-pass-prediction/blob/master/feature-list.md.
[3] https://github.com/henglicad/mlsa18-pass-prediction/blob/master/features.tsv.

- **Sender position features.** This dimension of features capture the position of the sender on the field, such as the sender's distance to the other team's goal. We choose this dimension of features because players have different passing strategies at different positions, for example, players may pass the ball more conservatively in their own half.
- **Candidate receiver position features.** This dimension of features capture the position of a candidate receiver, such as the candidate receiver's distance to the sender. Senders always consider candidate receivers' positions when they decide to whom to pass the ball.
- **Passing path features.** This dimension of features measure the quality of a passing path (i.e., the path from the sender to a candidate receiver), such as the passing angle. The quality of a passing path can predict the outcome (success/failure) of a pass.
- **Team position features.** This dimension of features capture the overall position of the team in control of the ball, such as the front line of the team. Team position might also impact the passing strategy, for example, a defensive team position might be more likely to pass the ball forwards.
- **Game state features.** This dimension of features capture the state of the whole game, such as the time when the sender passes the ball. **We do not use the time when the receiver receives the ball as a feature in our model, as it exposes information about the actual pass (e.g., pass duration).**

3.2   Modeling Approach

We formulate the task of predicting the receiver of a football pass as a learning to rank problem [8]. For each pass, our learning to rank model outputs a ranked list of the candidate receivers. A good model should rank the correct receiver at the front of the ranked list. LambdaRank [2] is a general and widely-used learning to rank framework. LambdaRank relies on underlying regression models to provide ranking predictions. **LambdaMART** [2] is the boosting tree version of LambdaRank. It relies on a gradient boosting decision tree (GBDT) [5] to provide ranking predictions. There are quite a few effective implementations of LambdaMART, such as XGBoost and pGBRT, which usually achieve state-of-the-art performance in learning to rank tasks.

In this work, we use an efficient implementation of LambdaMART, **Light-GBM** [7], which speeds up the training time of conventional LambdaMART implementations (e.g., XGBoost and pGBRT) by up to 20 times while achieving almost the same accuracy. We use an open source implementation of LightGBM that is contributed by Microsoft[4].

We use a 10-fold cross-validation to build and evaluate our model. The passes data is randomly partitioned into 10 subsets of roughly equal size. We build our model using nine subsets (i.e., the model building data) and evaluate the

---

[4] https://github.com/Microsoft/LightGBM.

performance of our model on the held-out subset (i.e., the testing data). The process repeats 10 times until all subsets are used as testing data once.

In each fold, we further split the model building data into the training data and validation data. We train the model on the training data and use the validation data to tune the hyper-parameters of the model. We do a grid search to get the top three sets of hyper-parameter values according to the performance of the model on the validation data. Then, we build three models with these three set of hyper-parameters using the training data. We apply these three models on the testing data and get three sets of results. We then average the results for each receiver candidate and use the averaged results to rank the receiver candidates. We find that with such an ensemble modeling approach, the accuracy of our model improves up to 2%.

3.3 Baseline Models

In order to evaluate the performance of our LightGBM ranking model, we compare it with several baseline models. As discussed in Sect. 2, 75% of the passes are within 20 m (i.e., short passes), and 62% of the passes are forward passes. Therefore, we derive baseline models that tend to select the nearest teammates and the teammates in the forward direction as the receiver.

- **The RandomGuess model** selects the receiver of a pass by a random guess. It randomly ranks the candidate receivers.
- **The NearestPass model** selects the nearest teammate of the sender as the top candidate receiver. It ranks the candidate receivers by their distance to the sender, from the teammates of the sender to the opponents, and then from the closest to the furthest.
- **The NearestForwardPass model** selects the nearest teammate of the sender that is in the forward direction (relative to the sender) as the top candidate receiver. It ranks the candidate receivers by their relative position to the sender, from the teammates of the sender to the opponents, then from the players in the forward direction to the players in the backward direction, and finally from the closest to the furthest.

3.4 Evaluation Approaches

We use **top-N accuracy** and **mean reciprocal rank (MRR)** to measure the performance of our model. Top-N accuracy measures the accuracy of the model's top-N recommendations, i.e., the probability that the correct receiver of a pass appears in the top-N receiver candidates that are predicted by the model. For example, top-1 accuracy measures the probability that the correct receiver of a pass is the first player in the predicted list of receiver candidates.

Reciprocal rank is the inverse of the rank of the correct receiver of a pass in an ranked list of candidate receivers predicted by the model. MRR [3] is the average of the reciprocal ranks over a sample of passes P:

Table 3. The accuracy of our model for predicting the receiver of a pass (excluding false passes).

	Back-field	Middle-field	Front-field	Overall
Top-1 accuracy	53%	46%	55%	50%
Top-3 accuracy	84%	81%	91%	84%
Top-5 accuracy	93%	93%	97%	94%
MRR	0.70	0.66	0.73	0.68

$$\text{MRR} = \frac{1}{|P|} \sum_{p=1}^{|P|} \frac{1}{\text{rank}_p} \qquad (1)$$

where rank_p is the rank of the correct receiver for the pth pass. The reciprocal value of MRR corresponds to the harmonic mean of the ranks. MRR ranges from 0 to 1, the larger the better. While top-N accuracy captures how likely the correct receiver appears in the top-N predicted receivers, MRR captures the average rank of the correct receiver in the predicted list of receiver candidates.

As discussed in Sect. 3.2, we use a 10-fold cross-validation to build and evaluate our model. Therefore, we use a mean top-N accuracy and MRR across the 10 folds in Sect. 4.

3.5   Feature Importance

In order to understand the importance of the features in our model, we use the feature importance scores that are automatically provided by a trained LightGBM model. Gradient boosting decision trees (e.g., LightGBM) provide a straightforward way to retrieve the importance scores of each feature [4].

After the boosting decision trees are constructed, for each decision tree, the importance of a feature is calculated by the amount that the feature improves the performance measure at its split point (i.e., split gains). The importance of each feature is then accumulated across all of the decisions trees in the model.

4   Results

This section discusses the answers to our research questions.

4.1   RQ1: How Well Can We Model the Receiver of a Pass?

Our model can predict the receiver of a pass with a top-1, top-3 and top-5 accuracy of 50%, 84%, and 94%, respectively, when we exclude false passes (i.e., passes to the other team). Table 3 shows the performance of our model when we exclude false passes. The "Back-field", "Middle-field",

Table 4. Comparing the accuracy of our model with baseline models (excluding false passes).

	LightGBM	RandomGuess	NearestPass	NearestForwardPass
Top-1 accuracy	**50%**	10%	33%	27%
Top-3 accuracy	**84%**	30%	70%	54%
Top-5 accuracy	**94%**	50%	86%	71%
MRR	**0.68**	0.29	0.55	0.47

Table 5. The accuracy of our model for predicting the receiver of a pass (considering all passes including passes to the other team).

	Back-field	Middle-field	Front-field	Overall
Top-1 accuracy	45%	38%	43%	41%
Top-3 accuracy	72%	68%	72%	70%
Top-5 accuracy	82%	80%	83%	81%
MRR	0.61	0.56	0.60	0.58

Table 6. Comparing the accuracy of our model with baseline models (considering all passes including passes to the other team).

	LightGBM	RandomGuess	NearestPass	NearestForwardPass
Top-1 accuracy	**41%**	5%	27%	23%
Top-3 accuracy	**70%**	14%	58%	45%
Top-5 accuracy	**81%**	24%	71%	59%
MRR	**0.58**	0.17	0.47	0.40

"Front-field" and "Overall" columns show the performance of our model for back-field passes, middle-field passes, front-field passes and all passes, respectively. A top-3 accuracy of 84% for all passes means that the actual receiver of a pass has a 84% chance to appear in our top-3 predicted candidates. The MRR value for all passes is 0.68, which means on average, the correct receiver is ranked 1.5th (i.e., 1/0.68) out of 10 or less receiver candidates (i.e., all teammates of the sender).

Our model can predict the receiver of a pass with a top-1, top-3 and top-5 accuracy of 41%, 70%, and 81%, respectively, when we consider all passes. Table 5 shows the performance of our model when we consider all passes (including false passes). The performance of our model decreases when we consider false passes (i.e., passes to the other team). False passes are very difficult to predict because it is not the sender player's intention to pass the ball to the other team. The MRR value for all passes is 0.58, which means the correct receiver is averagely ranked 1.7th (i.e., 1/0.58) out of all 21 or less candidate receivers (i.e., all players excluding the sender).

Table 7. The combined importance of each feature dimension.

Feature dimension	Combined feature importance
Candidate receiver position	6723
Team position	3493
Sender position	2688
Passing path	1753
Game state	343

Our model performs better for back-field and front-field passes, while performing worse for middle-field passes. Tables 3 and 5 also shows the performance of our model for back-field, middle-field and front-field passes, separately. Surprisingly, the performance of our model is the worst for middle-field passes. A player in the middle area may have more passing options, thereby increasing the difficulty to predict the right receivers.

Our model perform better than the baseline models. Tables 4 and 6 compare the performance of our LightGBM model with the RandomGuess, NearestPass, and NearestForwardPass models which are described in Sect. 3. Our LightGBM model consistently show much better performance than the three baseline models in terms of the top-N accuracy and MRR. The NearestPass model, which tends to pass the ball to the nearest teammates, achieve a better performance than the NearestForwardPass, which tends to pass the ball to the nearest teammates in the forward direction relative to the sender. Both of the NearestPass and NearestForwardPass baseline models achieve a much better performance than randomly guessing the receiver of a pass.

> Our model can predict the receiver of a pass with a top-1, top-3 and top-5 accuracy of 50%, 84%, and 94%, respectively, when we exclude false passes, outperforming three baseline models. Our model performs better when the sender of a pass is in the back or front area of the field.

4.2   RQ2: What Are the Important Factors that Explain the Receiver of a Pass?

The features that capture the candidate receivers' positions play the most important roles in explaining the receiver of a pass. Table 7 shows the combined importance of each feature dimension in our model. The combined importance is a sum of the importance scores of all the individual features in a dimension. The features from the dimension of candidate receiver position, which capture a candidate receiver's position on the pitch and his/her position relative to the teammates and opponents, have the biggest combined importance score in our model. The team position features, which captures the overall position of

Table 8. The ten most important features and their importance scores.

Dimension	Feature	Importance	Description
Receiver position[a]	receiver_closest_ opponent_dist	715	The distance between a candidate receiver and his/her closest opponent
Receiver position	norm_receiver_ sender_x_diff	660	Normalized x-axis difference between a candidate receiver and the sender
Receiver position	abs_y_diff	653	The absolute value of the y-axis difference between the sender and a candidate receiver
Receiver position	distance	616	A candidate receiver's distance to the sender
Receiver position	receiver_closest_ opponent_to_ sender_dist	602	The distance between the sender and a candidate receiver's closest opponent
Receiver position	receiver_closest_3_ opponents_dist	558	The average distance between a candidate receiver and his/her three closest opponents
Receiver position	receiver_to_center_ distance	521	The distance between a candidate receiver and the center of the pitch
Receiver position	receiver_closest_3_ teammates_dist	508	The average distance between a candidate receiver and his/her three closest teammates
Sender position	sender_closest_ opponent_dist	498	The distance between the sender and his/her closest opponent
Passing path	min_pass_angle	467	Pass angles are the angles between the line from the sender to a candidate receiver (i.e., the pass line) and the lines from the sender to the opponents along the pass line. The min_pass_angle is the minimum pass angle for a pass line

[a] "Receiver position" is short for "candidate receiver position".

the team in control of the ball, are the second important dimension in explaining the receiver of a pass. The third important feature dimension (i.e., sender position) captures the sender's position on the pitch and his/her position relative to the teammates and opponents. The passing path features, which captures the characteristics of a passing path (i.e., the path from the sender to a candidate receiver), also play a significant role in explaining the receiver of a pass.

The most important features capture the candidate receivers' positions relative to the sender and the opponents. Table 8 lists the top ten features that are most important in our model and their respective importance scores. A full list of our features' important scores is available online (See footnote 4). Among the top 10 important features, there are eight features from the

dimension of candidate receiver position. All of the top six features are from the
dimension of candidate receiver position, among which three features capture a
candidate receiver's relative position to the sender, and the other three features
capture a candidate receiver's relative position to the components. The other two
features from the dimension of candidate receiver position capture a candidate
receiver's position on the pitch and his/her relative position to the teammates,
respectively. Among the top 10 important features, there are also one from the
sender position dimension (i.e., the sender_closest_opponent_dist feature), and
one from the passing path dimension (i.e., the min_pass_angle feature).

> The features that capture the positions of the candidate receivers, in partic-
> ular, relative to the sender and the opponents, play the most important roles
> in explaining the receiver of a pass.

5 Conclusions

This work proposes a novel approach to predict the receivers of football passes.
We analyze a dataset containing 12,124 passes from 14 real-world football games
and discuss players' passing characteristics. We find that players present different
passing characteristics in different areas of the field. We then extract 54 features
along five dimensions and build a LightGBM model to predict the receiver of a
pass. Our model achieves a top-1, top-3, and top-5 accuracy of 50%, 84%, and
94%, respectively, when we exclude false passes. Our model outperforms three
baseline models that we use to rank the candidate receivers of a pass. We find
that the features that capture the positions of the candidate receivers play the
most important roles in explaining the receiver of a pass. We believe that our
approaches and findings can help football practitioners better understand the
factors that impact the receiver of a pass and make informed tactical decisions.

References

1. Ali, A.: Measuring soccer skill performance: a review. Scand. J. Med. Sci. Sports
   21(2), 170–183 (2011)
2. Burges, C.J.: From ranknet to lambdarank to lambdamart: an overview. Learning
   11(23–581), 81 (2010)
3. Craswell, N.: Mean reciprocal rank. In: Liu, L., Özsu, M.T. (eds.) Encyclopedia of
   Database Systems, p. 1703. Springer, Boston (2009). https://doi.org/10.1007/978-
   0-387-39940-9_488
4. Friedman, J., Hastie, T., Tibshirani, R.: The Elements of Statistical Learning. SSS,
   vol. 1. Springer, New York (2009). https://doi.org/10.1007/978-0-387-84858-7
5. Friedman, J.H.: Greedy function approximation: a gradient boosting machine. Ann.
   Stat. **29**(5), 1189–1232 (2001)
6. Hughes, M., Franks, I.: Analysis of passing sequences, shots and goals in soccer. J.
   Sports Sci. **23**(5), 509–514 (2005)

7. Ke, G., et al.: LightGBM: a highly efficient gradient boosting decision tree. In: Guyon, I., et al. (eds.) Advances in Neural Information Processing Systems 30, pp. 3146–3154 (2017)
8. Liu, T.Y., et al.: Learning to rank for information retrieval. Found. Trends® Inf. Retrieval **3**(3), 225–331 (2009)
9. Reep, C., Benjamin, B.: Skill and chance in association football. J. R. Stat. Soc. Ser. A (General) **131**(4), 581–585 (1968)

Author Index

Printed in the United States
By Bookmasters